8th Young Researcher Meeting – Cagliari 2017

Journal of Physics: Conference Series Volume 956

Cagliari, Italy
29 May - 1 June 2017

Editor:

Fabio Agostini

ISBN: 978-1-5108-5756-8
ISSN: 1742-6588

Printed from e-media with permission by:

Curran Associates, Inc.
57 Morehouse Lane
Red Hook, NY 12571

Printed by Curran Associates, Inc. (2018)

For permission requests, please contact the Institute of Physics
at the address below.

Institute of Physics
Dirac House, Temple Back
Bristol BS1 6BE UK

Phone: 44 1 17 929 7481
Fax: 44 1 17 920 0979

techtracking@iop.org

Additional copies of this publication are available from:

Curran Associates, Inc.
57 Morehouse Lane
Red Hook, NY 12571 USA
Phone: 845-758-0400
Fax: 845-758-2633
Email: curran@proceedings.com
Web: www.proceedings.com

Table of contents

Volume 956

8th Young Researcher Meeting

29 May to 1 June 2017, Cagliari, Italy

Accepted papers received: 9 January 2018
Published online: 1 February 2018

Preface

Papers

Astrophysics and Cosmology

Condensed Matter Physics

IOP Conf. Series: Journal of Physics: Conf. Series **956** (2017) 011001 doi:10.1088/1742-6596/956/1/011001

Young Researcher Meeting, Cagliari 2017

Fabio Agostini, Claudia Antolini, Pierpaolo Cacciotto, Matteo Cadeddu, Ester Casula, Giordano Cattani, Alessandro Cultrera, Marco Di Stefano, Carlo Flore, Edgardo Franzin, Luca Giovannelli, Elisa Incani, Francesca Loi, Marina Migliaccio, Lorenzo Pagnanini, Fabrizio Piacentini, Davide Pietrobon, Emanuela Pusceddu, Valerio Sarritzu, Matteo Serra, Francesco Stellato, Matteo Tuveri

YRM YOUNG RESEARCHER MEETING

IPN INTERNATIONAL PHYSICISTS NETWORK

PREFACE

Held annually since 2009, the Young Researcher Meeting (iphysnet.com/wp/yrm) gathers together students, post-doc fellows, and young researchers eager to discuss the current progress and developments in various areas of physics.

The dissemination of results and sharing of knowledge are indeed activities that should be valorised since the early stages of anyone's scientific career. This has long been the spirit of the conference, and it has a deep influence on its format. In fact, the Young Researcher Meeting aims to create strong bonds among the attendees and to promote collaboration and cross-fertilisation among different research areas. It does so in an informal set-up, where students and young researchers introduce their work to their peers in brief presentations and a lively poster session. Questions, discussions, and interaction come naturally and this has brought ever-increasing success to the conference, making it an unmissable appointment for young researchers.

The conference has been hosted in various locations around Italy by several major institutions throughout the years. The 8th edition was held at the Cittadella dei Musei in Cagliari, from May 29th to June 1st 2017, and it was jointly organised by the freshly established International Physicists Network association (iphysnet.com) and the Department of Physics of the University of Cagliari.

IOP Conf. Series: Journal of Physics: Conf. Series **956** (2017) 011001 doi:10.1088/1742-6596/956/1/011001

This edition rounded up over 60 participants belonging to universities and research centres in Europe, Asia, and South America. The plenary talk sessions covered areas such as Astrophysics and Cosmology, Biomedical and Particle Physics, Condensed Matter and Theoretical Physics — topics that have been deepened and widened in the more interactive and informal *poster & wine* session.

Moreover, the programme included a guided tour to the Sardinia Radio Telescope; a *lectio magistralis* by Prof. Cristiano Galbiati on dark matter and neutrinoless double beta decays; and invited talks by Dr. Carlos Mejia-Monasterio from the Technical University of Madrid, Spain, and Dr. Paolo Pani from the University of Rome "Sapienza", who presented cutting edge research in their respective fields in an appealing and enjoyable way. During the conference, the first meeting of the International Physicists Network took place, an Italian non-profit association that aims to connect young physicists across the world and presides the organisation of the Young Researcher Meeting. This edition also featured a pleasant social dinner in a local restaurant.

This volume consists of a selection of the contributions presented at the conference as either talks or posters. These cover topics in Astrophysics and Cosmology, Biomedical and Particle Physics, Condensed Matter and Theoretical Physics.

Organising and Editorial Committee

Fabio Agostini
(fabioagostini31@gmail.com)
Akka Technologies, Rome, Italy

Claudia Antolini
(claudiaantolini@cambridgesceincecentre.com)
Cambridge Science Centre, Cambridge, UK

Pierpaolo Cacciotto
(p.cacciotto@outlook.com)

Matteo Cadeddu
(matteo.cadeddu@ca.infn.it)
Dipartimento di Fisica, Università degli Studi di Cagliari & INFN, Sezione di Cagliari, Cagliari, Italy

Ester Casula
(ester.anna.rita.casula@cern.ch)
INFN Sezione di Cagliari, Cagliari, Italy

Giordano Cattani
(giordano.cattani@asml.com)
ASML, Eindhoven, Netherlands

Alessandro Cultrera
(a.cultrera@inrim.it)
INRiM, Turin, Italy

Marco Di Stefano
(marco.distefano@cnag.crg.eu)
Centre Nacional d'Anàlisi Genòmica - Centre de Regulació Genòmica (CNAG-CRG), Barcelona, Spain

Carlo Flore
(carlo.flore@ca.infn.it)
Dipartimento di Fisica, Università degli Studi di Cagliari & INFN, Sezione di Cagliari, Cagliari, Italy

Edgardo Franzin
(edgardo.franzin@ca.infn.it)
Dipartimento di Fisica, Università degli Studi di Cagliari & INFN, Sezione di Cagliari, Cagliari, Italy

Luca Giovannelli
(luca.giovannelli@roma2.infn.it)
Università degli studi di Roma "Tor Vergata", Rome, Italy

Elisa Incani
(elisa.incani@gmail.com)
INFN, Sezione di Cagliari, Cagliari, Italy

Francesca Loi
(floi@oa-cagliari.inaf.it)
Dipartimento di Fisica, Università degli Studi di Cagliari & INAF-Osservatorio Astronomico di Cagliari, Cagliari, Italy

Marina Migliaccio
(marina.migliaccio@ssdc.asi.it)
Space Science Data Center - ASI, Rome, Italy

Lorenzo Pagnanini
(lorenzo.pagnanini@gssi.infn.it)
Gran Sasso Science Institute, L'Aquila, Italy

Fabrizio Piacentini
(f.piacentini@inrim.it)
INRiM, Turin, Italy

Davide Pietrobon
(daddeptr@gmail.com)
HERE Technologies, Berkeley, USA

Emanuela Pusceddu
(emanuela.pusceddu@gmail.com)
Cagliari, Italy

Valerio Sarritzu
(valerio.sarritzu@dsf.unica.it)
Dipartimento di Fisica, Università degli Studi di Cagliari, Cagliari, Italy

Matteo Serra
(mserra@fbk.eu)
Fondazione Bruno Kessler, Trento, Italy

Francesco Stellato
(francesco.stellato@roma2.infn.it)
INFN Roma Tor Vergata, Rome, Italy

Matteo Tuveri
(matteo.tuveri@ca.infn.it)
Dipartimento di Fisica, Università degli Studi di Cagliari & INFN, Sezione di Cagliari, Cagliari, Italy

Acknowledgments

The organisers of the 8th Young Researcher Meeting, held in Cagliari, would like to thank all the authors for their scientific contributions and all the people who made this conference smooth and pleasant.

We thank the University of Cagliari for hosting the conference, the Physics Department, the Osservatorio Astronomico di Cagliari (OAC), the Istituto Nazionale di AstroFisica (INAF), the Istituto Nazionale di Fisica Nucleare (INFN), the Istituto di Scienze dell'Atmosfera e del Clima (CNR-ISAC) for financial support.

We acknowledge support from L'Apesarda Aresu, Sa Defenza and Salumificio Parteolla.

We are grateful to Prof. C. Galbiati for his enthusiasm in accepting to deliver the *lectio magistralis*, to Dr. C. Mejia-Monasterio and Dr P. Pani for sharing their knowledge and experience as invited speakers, and to Dr. A. Possenti, Dr. M. Bacchetti, Dr. M. Murgia and Dr. M. Burgay for having organised and hosted the visit to the Sardinia Radio Telescope (http://www.srt.inaf.it/).

The event was broadcast live on the conference website, and the recordings are available on the YRM Youtube channel at http://www.youtube.com/c/YoungResearcherMeeting.

IOP Conf. Series: Journal of Physics: Conf. Series **956** (2017) 011001 doi:10.1088/1742-6596/956/1/011001

IOP Conf. Series: Journal of Physics: Conf. Series **956** (2017) 011002 doi:10.1088/1742-6596/956/1/011002

Peer review statement

All papers published in this volume of *Journal of Physics: Conference Series* have been peer reviewed through processes administered by the proceedings Editors. Reviews were conducted by expert referees to the professional and scientific standards expected of a proceedings journal published by IOP Publishing.

IOP Conf. Series: Journal of Physics: Conf. Series **956** (2017) 012001 doi:10.1088/1742-6596/956/1/012001

ISW-galaxy cross-correlation in K-mouflage

G Benevento[1,2], **N Bartolo**[1,2,3], **M Liguori**[1,2,3]

[1] Dipartimento di Fisica e Astronomia "G. Galilei", Università degli Studi di Padova, via Marzolo 8, I-35131, Padova, Italy
[2] INFN, Sezione di Padova, via Marzolo 8, I-35131, Padova, Italy
[3] INAF-Osservatorio Astronomico di Padova, Vicolo dellOsservatorio 5, I-35122 Padova, Italy

E-mail: `benevent@pd.infn.it`

Abstract. Cross-correlations between the cosmic microwave background and the galaxy distribution can probe the linear growth rate of cosmic structures, thus providing a powerful tool to investigate different Dark Energy and Modified Gravity models. We explore the possibility of using this observable to probe a particular class of Modified Gravity models, called K-mouflage.

1. Introduction

In a Universe undergoing accelerating expansion, the gravitational potentials associated to large scale structures decay. A photon of the Cosmic Microwave Background (CMB) travelling through a decaying potential will experience a net change in energy [1], which leads to a secondary anisotropy in the CMB temperature distribution. Such particular gravitational redshift effect is called Integrated Sachs Wolfe (ISW) [2], as it has to be integrated along the line of sight. This effect can be used to discriminate between different Dark Energy (DE) and Modified Gravity (MG) scenarios [3, 4, 5]. The ISW signal is not detectable using only CMB temperature data, due to the noise arising from other sources of anisotropies on large angular scales and from the cosmic variance. Cross-correlation between the CMB and a tracer of the gravitational potential, like the projected galaxy number counts, can extract the ISW signal, making it detectable with high significance using the upcoming galaxy surveys [5].

In this paper we investigate the power of this probe to discriminate between the Λ-Cold Dark Matter model (ΛCDM) and the K-mouflage models of modified gravity [6]. K-mouflage models, that we present in Section 2, predict a considerable deviation from the ΛCDM model in the growth rate of cosmic structures [7], this feature makes them suitable to be tested through cross-correlation between CMB and galaxies. In Section 3 we perform a cross-correlation analysis for two different K-mouflage models, with the aim of forecasting their detectability using two upcoming cosmological surveys: Euclid [8] and LSST [9]. We draw our conclusions in Section 4.

2. Growth of linear perturbations in K-mouflage models

K-mouflage models assume the presence of an additional scalar field (φ) non-minimally coupled to the space-time metric ($g_{\mu\nu}$) [6]. The action in the Einstein frame can be expressed in the following form:

$$S = \int \mathrm{d}^4x\sqrt{-g}\left[\frac{M_{Pl}^2}{2}R + \mathcal{M}^4K(\chi)\right] + \int \mathrm{d}^4x\sqrt{-\tilde{g}}\mathcal{L}_m(\psi_m^{(i)}, \tilde{g}_{\mu\nu}), \quad (1)$$

	Kinetic term	Coupling function	β	m	K_0
Model 1	$-1+\chi+K_0\chi^m$	$A=e^{\beta\varphi\sqrt{8\pi G}}$	0.3	3	1
Model 2	$-1+\chi+K_0\chi^m$	$A=e^{\beta\varphi\sqrt{8\pi G}}$	0.3	3	-5

Table 1. Definition of the K-mouflage models we adopt in our analysis.

$$\chi = -\frac{1}{2\mathcal{M}^4} g_{\mu\nu}\partial^\mu\varphi\partial^\nu\varphi, \qquad \tilde{g}_{\mu\nu} = A^2(\varphi) g_{\mu\nu}, \tag{2}$$

where M_{Pl} is the Planck mass, M^4 is the energy scale of the field φ and $\mathcal{L}_m$ is the Lagrangian of the matter fields $\psi_m^{(i)}$. This action involves two metrics, the Einstein-frame metric $g_{\mu\nu}$ and the Jordan-frame metric $\tilde{g}_{\mu\nu}$, related to the former by the conformal transformation in Eq. (2). K-mouflage models are characterized by a scalar field Lagrangian with a non-canonical kinetic term $K(\chi)$. The form of the kinetic function and of the coupling function needs to be specified and different choices are possible. Here, we consider two simple models, for which the kinetic and coupling functions are given in Table 1.
The cosmological perturbation theory for K-mouflage models has been studied in [7]. We report here only the main results of the perturbative treatment at the linear level.
The perturbed Robertson-Walker metric in the Einstein frame, using the Newtonian conformal gauge, takes the same form as in ΛCDM:

$$\mathrm{d}s^2 = a^2(\eta)\left[-\left(1+2\Phi(\vec{x},\eta)\right)\mathrm{d}\eta^2 + \left(1-2\Psi(\vec{x},\eta)\right)\delta_{ij}\mathrm{d}x^i\mathrm{d}x^j\right], \tag{3}$$

where η is the conformal time, δ_{ij} is the Kronecker delta and a is the scale factor of the universe. The difference between the two scalar perturbations Φ and Ψ is negligible at the linear perturbative order in the Einstein frame [7], thus we consider from now on $\Phi = \Psi = \Psi_N$. We assume small fluctuations in the densities ($\delta\rho_i = \rho_i - \bar{\rho}_i$) and in the scalar field ($\delta\varphi = \varphi - \bar{\varphi}$). If we Fourier transform the density field, we find that during the matter-dominated era the linear k-modes evolve independently in K-mouflage similarly to the ΛCDM case [7]. We can parametrize the growth of matter perturbations in the following way:

$$\delta(\vec{k},\eta) = \delta_0(\vec{k})D(\eta), \tag{4}$$

where $\delta(\vec{k},\eta)$ is the Fourier transform of the matter density contrast $\delta\rho/\bar{\rho}$, $D(\eta)$ is the linear growth factor of perturbations and δ_0 refers to the present epoch.
Analogously to the ΛCDM case, combining the linearised continuity equation and Euler equation on sub-horizon scales, we obtain the evolution equation for the growth factor in K-mouflage [7]

$$\frac{\mathrm{d}^2D}{\mathrm{d}\eta^2} + aH(1+\epsilon_2)\frac{\mathrm{d}D}{\mathrm{d}\eta} - a^2H^2\frac{3}{2}\Omega_m(1+\epsilon_1)D = 0, \tag{5}$$

$$\epsilon_1(\eta) \equiv \bar{A} - 1 + \frac{2\bar{A}\beta^2}{\bar{K}'}, \qquad \epsilon_2 \equiv \frac{\mathrm{dln}\bar{A}}{\mathrm{dln}a}, \tag{6}$$

where $\bar{A}$ is the space-averaged coupling function, $\bar{K}'$ is the space-averaged derivative of $K(\chi)$ w.r.t. χ, Ω_m is the time-dependent matter density parameter in K-mouflage $\Omega_m = \bar{\rho}_m/(3M_{Pl}^2H^2)$ and H is the expansion rate of the universe. The coefficients ϵ_1 and ϵ_2 encode the deviation of K-mouflage from the ΛCDM growth of perturbations.
The growing mode of the gravitational potential is related to the growth factor by the Poisson equation. In the ΛCDM model, this relation simply reads $D_{\Psi_N} = \frac{D}{a}$, whereas in K-mouflage we can write the linear growing mode D_{Ψ_N} of the gravitational potential as:

$$D_{\Psi_N} = \bar{A}\frac{D}{a}. \tag{7}$$

3. ISW-galaxy cross-correlation

In a Dark Energy dominated universe the gravitational potential varies with time unlike the case of a matter dominated universe [3]. This leaves an imprint on the CMB power spectrum. This phenomenon, the so-called late Integrated Sachs-Wolfe (late ISW) effect, can play a primary role in discriminating between different dynamical DE or MG models, since it is sensitive to the evolution of the gravitational potential, which is encoded in the growth factor.

Solving Eq. (5), as done e.g. in [7], shows that K-mouflage models predict substantial deviations in the growth factor, as well as in its derivative, w.r.t. ΛCDM. In particular the gravitational potential in K-mouflage is expected to evolve even during the matter-dominated era, at redshift $z > 2$, when it is constant in the standard cosmological scenario. This peculiar growth of matter perturbations produces a signature in the late ISW anisotropies, which can be used as an observational probe for this class of models.

The cross-correlation between galaxy distribution and CMB anisotropies can enhance the late ISW signal, making it detectable, thus it can be successfully used to constrain K-mouflage models. The key quantity to describe the amount of cross-correlation between the CMB and the galaxy number counts, due to the late ISW effect, is the cross-correlation power spectrum. In analogy with the definition of the CMB angular power spectrum, we can define the ISW-galaxy cross-power spectrum as the ensemble average of the product of the harmonic coefficients a_{lm} obtained expanding the galaxy count fluctuations and the CMB temperature fluctuations in spherical harmonics:

$$\left\langle a_{lm}^{ISW} \left(a_{l'm'}^{gal} \right)^* \right\rangle \equiv C_l^{ISW-gal} \delta_{ll'} \delta_{mm'} . \tag{8}$$

As shown in [4], we can express $C_l^{ISW-gal}$ as follows:

$$C_l^{ISW-gal} = \frac{2}{\pi} \int \mathrm{d}k \; k^2 P(k) I_l^{ISW} I_l^{gal}, \tag{9}$$

$$I_l^{gal}(k) = \int_0^{\infty} \mathrm{d}z \; \frac{\mathrm{d}N}{\mathrm{d}z} b(z) D(z) j_l(kr(z)), \tag{10}$$

$$I_l^{ISW}(k) = 3\Omega_m^{(0)} \left(\frac{H_0}{k} \right)^2 \int_0^{z_{rec}} \mathrm{d}z \; \frac{\mathrm{d}}{\mathrm{d}z} \left[D_{\Psi_N}(z) \right] j_l(kr(z)), \tag{11}$$

where $P(k)$ is the matter power spectrum, $\frac{\mathrm{d}N}{\mathrm{d}z}$ is the galaxy distribution as function of the redshift z and $b(z)$ is the bias parameter which relates the density contrast of the dark matter to the density contrast of the galaxy distribution. j_l are the spherical Bessel functions which depend on the comoving distance $r(z)$. $\Omega_m^{(0)}$ and H_0 are respectively the present matter density parameter and the Hubble constant. The two integrals I_l^{gal} and I_l^{ISW} account respectively for the contributions of the galaxy number counts and ISW temperature anisotropies. In the last equation the linear growing mode of the gravitational potential D_{Ψ_N} assumes the form given by Eq. (7) while it is simply equal to $(1+z)D$ for the ΛCDM model.

The computation of the cross-correlation power spectrum can be simplified by applying the Limber approximation, valid for sufficiently high multipole moments l, see [10]. Using the Limber approximation, we can finally rewrite the cross-power spectrum as an integral over z only:

$$C_l^{ISW-gal} \approx \frac{-3H_0^2\Omega_m^{(0)}}{(l+1/2)^2} \int \mathrm{d}z \; b(z) \frac{\mathrm{d}N}{\mathrm{d}z} \; D(z) \frac{\mathrm{d}}{\mathrm{d}z} [D_{\Psi_N}(z)] P\left[\frac{(l+1/2)}{r(z)} \right] . \tag{12}$$

3.1. Analysis

We now develop a cross-correlation analysis, comparing the predictions for different K-mouflage models with those of ΛCDM and estimating the significance of a possible future detection.

Extremely deep and wide surveys will gather an exceptional catalogue of sources in the next

	Euclid-like survey	LSST-like survey
Number density	$n_g = 30$ arcmin^{-2}	$n_g = 50$ arcmin^{-2}
Sky coverage	15.000 deg^2	20.000 deg^2
Redshift distribution	$\alpha = 2,\ \beta = 1.5,\ z_0 = 0.64$	$\alpha = 2,\ \beta = 1,\ z_0 = 0.5$
Median redshift	$z_m = 0.9$	$z_m = 1.34$
photo-z errors	$\sigma_z = 0.05(1+z)$	$\sigma_z = 0.02(1+z)$
galaxy bias	$b = \sqrt{1+z}$	$b = 1 + 0.84z$

Table 2. Properties of the upcoming surveys considered in this analysis. We consider photometric surveys, for Euclid we assume the same predictions given in [8], while for LSST we follow the forecast of [9].

few years (e.g. [8, 9]). In particular their high limiting magnitude (e.g. 24.5 for Euclid and 27.5 for LSST) will allow to map the galaxy distribution up to very high redshifts. This makes them ideal for a tomographic study of the ISW effect through CMB-galaxy cross-correlation. For our purpose we concentrate on two promising surveys, Euclid and LSST, that are optimized for CMB-galaxy cross-correlation as shown in [8, 9].
The main goal of this analysis is to forecast the possibility of investigating K-mouflage using the differences in the late ISW effect predicted by such theory, opening the way to more detailed studies. With this aim we want to compute the signal to noise ratio for a possible detection of the ISW-galaxy correlation, assuming the ΛCDM prediction as null hypothesis.
Following the standard procedure for similar forecasts [10, 11, 12], we bin the (predicted) redshift distribution of galaxies for Euclid and LSST and we then compute ISW-galaxy cross-correlation power spectra solving numerically Eq. (9) (or Eq. (12) for $l > 20$), for the K-mouflage models defined in Table 1 and for ΛCDM.
The redshift distribution of galaxies for the upcoming surveys Euclid and LSST, can be well approximated by the following analytic form [13]:

$$\frac{\mathrm{d}N}{\mathrm{d}z} = \frac{1}{\Gamma\left(\frac{\alpha+1}{\beta}\right)}\beta\frac{z^{\alpha}}{z_0^{\alpha+1}}\exp\left[-\left(\frac{z}{z_0}\right)^{\beta}\right]. \tag{13}$$

The above parametrization gives the normalized distribution of galaxies per steradian. The parameters α, β and z_0 are characteristic of each survey and determine the shape of the distribution as well as its median redshift. Their values are given in Table 2. To simulate a tomographic analysis we arbitrarily chose to divide the Euclid-like galaxy distribution in 3 redshift bins, and the LSST-like survey in 5 redshift bins, following the procedure of [13].
Assuming a Gaussian distribution, we can estimate the error associated to $C_l^{ISW-gal}$ in a given redshift bin for a single multipole l as [10]:

$$\sigma^2\left(C_l^{ISW-gal}\right) = \frac{1}{f_s(2l+1)}\left[\left(C_l^{ISW-gal}\right)^2 + \left(C_l^{gg} + \frac{1}{n_g}\right)\left(C_l^{TT}\right)\right], \tag{14}$$

where C_l^{TT} and C_l^{gg} are the auto-correlation power spectra of CMB temperature anisotropies and of galaxies number counts respectively. f_s is the fraction of sky covered by the survey, while n_g is the number density of galaxies per steradian in the considered redshift bin.
Assuming that different l multipoles are uncorrelated, we can compute the cumulative squared signal to noise ratio of the deviation between K-mouflage models and the ΛCDM as:

$$\left(\frac{S}{N}\right)^2 = \sum_l \frac{\left[\left(C_l^{ISW-gal}\right)_{Km} - \left(C_l^{ISW-gal}\right)_{\Lambda CDM}\right]^2}{\sigma^2}, \quad \sigma^2 = \sigma^2_{Km} + \sigma^2_{\Lambda CDM}. \tag{15}$$

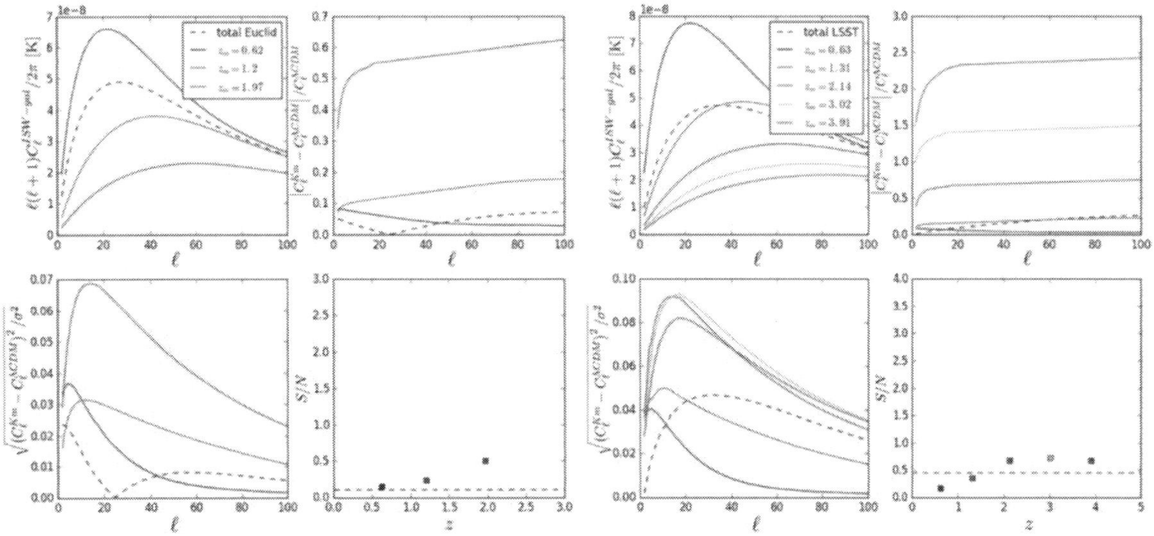

Figure 1. ISW-galaxies cross-correlation power spectra and S/N for *Model 1* of Table 1 as expected to be measured by Euclid and LSST. The four panels on the left represent quantities computed using the Euclid-like survey divided in 3 redshift bins, while the four panels on the right have been obtained considering the LSST-like survey divided in 5 redshift bins. Each bin is labelled with its median redshift z_m.

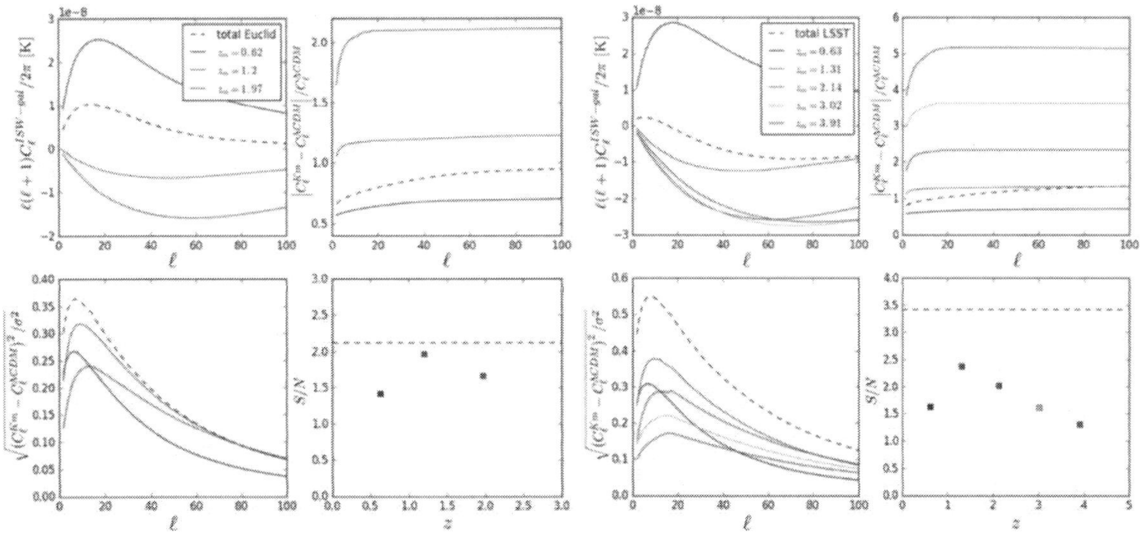

Figure 2. ISW-galaxies cross-correlation power spectra and S/N for *Model 2* of Table 1. The same quantities computed for *Model 1* are shown.

In Fig. 1 and Fig. 2 we show our results for the cross-correlation power spectra (adopting the compact notation $C_l^{Km} = \left(C_l^{ISW-gal}\right)_{Km}$ and $C_l^{\Lambda CDM} = \left(C_l^{ISW-gal}\right)_{\Lambda CDM}$) computed for the K-mouflage models of Table 1, their relative difference w.r.t. the ΛCDM prediction $\left|C_l^{Km} - C_l^{\Lambda CDM}\right| / C_l^{\Lambda CDM}$, the signal to noise ratio per multipole $\left|C_l^{Km} - C_l^{\Lambda CDM}\right| / \sigma$, and the total signal to noise ratio obtained using Eq. (15).

The signal to noise ratio per multipole peaks at $l \sim 10-60$: at low multipole orders the uncertainty on the power spectrum is dominated by the cosmic variance, while at large l the signal decreases due to the decay of the ISW on small angular scales. Our result for the cumulative S/N shows that the significance of the deviation of *Model 1* from ΛCDM would reach $\sim 0.5\sigma$ for the Euclid-like survey and $\sim 0.8\sigma$ using the LSST-like survey. For *Model 2* we obtain an higher significance, $\sim 2.2\sigma$ for the Euclid-like survey and $\sim 3.5\sigma$ for the LSST-like survey. This is due to the peculiar behaviour of $C_l^{ISW-gal}$ in *Model 2*, that shows an anti-correlation at high redshift. This feature makes the model significantly different from ΛCDM that predicts a positive correlation.
We observe that, especially for *Model 1*, we gain significance by splitting the galaxy distribution in redshift bins. The tomographic analysis allows to select sharp redshift intervals in which the correlation signal is high enough to overcome the increase in shot noise due to the reduced number of galaxies considered. Finally, we observe that the results obtained using the Euclid-like and the LSST-like survey are consistent with our expectations. In our forecast the LSST-like survey is supposed to cover a larger area of sky and reduce the shot noise in Eq. (14) observing an higher number of galaxies with respect to Euclid (see Table 2), these differences improve the detectability of K-mouflage models.

4. Conclusions

The measurement of the late-ISW signal through the cross-correlation between the CMB and the projected distribution of galaxies, observed by wide cosmological surveys, constitutes a promising method of investigation to discriminate between different DE and MG theories.
In this work we have computed the ISW-galaxy cross-correlation power spectra for two different K-mouflage scenarios. We found the striking result that *Model 2* predicts a negative ISW-galaxy cross-correlation, this could represent a smoking-gun signature for similar models. In order to forecast the detectability of K-mouflage models using future surveys like Euclid and LSST we performed an error analysis, computing the signal to noise ratio in several redshift bins. Our results confirm that the probing power of CMB-galaxies cross-correlations will increase with the upcoming generation of surveys, also thanks to the possibility to perform tomographic analysis.
A more general analysis of the K-mouflage phenomenology, including all the observable effects on CMB anisotropies, requires the full solution of the linearised Einstein-Boltzmann equations for this theory. This can be done e.g. using the EFTCAMB Einstein-Boltzmann solver [14]. Including K-mouflage models in the EFTCAMB code allows to study different observational signatures and to perform the comparison with data in a systematic way. The work based on this method is currently under development [15].

References

[1] Hu W and Dodelson S 2002 *ARA&A* **40** 171
[2] Sachs R K and Wolfe A M 1967 *ApJ* **147** 73
[3] Crittenden R, Boughn S and Turok N 1996 *Bull. Am. Astron. Soc.* **28** 1341
[4] Khosravi S, Mollazadeh A and Baghram S 2016 *JCAP* **9** 003
[5] Hu W and Scranton R 2004 *Phys. Rev. D* **70** 123002
[6] Brax P and Valageas P 2014 *Phys. Rev. D* **90** 023507
[7] Brax P and Valageas P 2014 *Phys. Rev. D* **90** 023508
[8] Amendola L, Appleby S, Avgoustidis A *et al* 2013 *Living Rev. Relativ.* **16** 6
[9] LSST Science Collaboration, Abell P A, Allison J *et al* 2009 preprint arXiv:0912.0201
[10] Schmidt F, Liguori M and Dodelson S 2007 *Phys. Rev. D* **76** 083518
[11] Bertacca D, Raccanelli A, Piattella O F *et al* 2011 *JCAP* **3** 039
[12] Douspis M, Castro P G, Caprini C and Aghanim N 2008 *Astron. Astrophys.* **485** 395
[13] de Putter R, Doré O and Das S 2014 *ApJ* **780** 185
[14] Raveri M, Hu B, Frusciante N and Silvestri A 2014 *Phys. Rev. D* **90** 043513
[15] Benevento G, Bartolo N, Brax P, Lazanu A, Liguori M, Raveri M and Valageas P in preparation.

IOP Conf. Series: Journal of Physics: Conf. Series **956** (2017) 012002 doi:10.1088/1742-6596/956/1/012002

Impact of polarized foregrounds on LSPE-SWIPE observations

A Buzzelli[1,2], M Migliaccio[3,2], G de Gasperis[2,4], P de Bernardis[1,5], S Masi[1,5] and N Vittorio[2,4]

[1] Dipartimento di Fisica, "Sapienza" Università di Roma, piazzale Aldo Moro 5, I-00185, Roma, Italy
[2] Sezione INFN Roma 2, via della Ricerca Scientifica 1, I-00133, Roma, Italy
[3] Space Science Data Center - ASI, Via del Politecnico snc, 00133, Roma, Italy
[4] Dipartimento di Fisica, Università di Roma "Tor Vergata", via della Ricerca Scientifica 1, I-00133, Roma, Italy
[5] Sezione INFN Roma 1, piazzale Aldo Moro 5, I-00185, Roma, Italy

Abstract. A first measurement of Cosmic Microwave Background B-mode polarization on large scales could provide a convincing confirmation of the existence of a primordial background of gravitational waves as predicted in the inflation scenario. A major obstacle to observe B-modes is represented by polarized foreground contamination from our Galaxy. In particular, thermal dust and synchrotron emission dominate over the primordial signal at high and low frequencies, respectively. Several experiments have been designed to observe B-mode polarization. Here, we focus on the forthcoming LSPE-SWIPE balloon experiment, devoted to the accurate observation of large scale CMB polarization, and present preliminary forecasts on the impact of foreground contamination on LSPE-SWIPE observations. Using the last release of Planck foreground maps as templates, we estimate the amplitude of dust and synchrotron emission in the sky region and at the frequency channels of interest for LSPE-SWIPE. Furthermore, we investigate the generation of polarization-optimized Galactic masks and we give preliminary indications on the requirements of component separation methods.

1. Introduction

The analysis of the Cosmic Microwave Background (CMB) temperature anisotropy pattern has provided tight limits on the early stage, evolution and content of the Universe [1]. The Planck experiment has given an ultimate dataset for intensity maps but just opened the observation of the CMB (linear) polarization pattern. The polarization component, usually described in terms of the so-called E and B-modes, can further constrain some crucial aspects of the cosmological model. E-mode polarization has been widely detected, while there has been no detection of primordial B-modes so far. B-mode polarization is uniquely sensitive to the tensor mode of primordial perturbations, i.e. the gravitational wave stochastic background whose existence is predicted by the inflationary paradigm. Moreover, B-modes could also help in constraining critical aspects of the reionization history of the Universe. Therefore, the detection of B-modes represents the new frontier of cosmological observations. The B-mode amplitude is usually expressed in terms of the tensor-to-scalar ratio, r, which gives the relative amplitude of tensor and scalar primordial perturbations. Planck, BICEP2 and Keck Array [2] constrained r down to 0.07 at 95% C.L., but we do not have accurate indications on its lower limit. However, theory

IOP Conf. Series: Journal of Physics: Conf. Series **956** (2017) 012002 doi:10.1088/1742-6596/956/1/012002

suggests $r > 10^{-3}$ [3]. A major obstacle to observe B-mode polarization is represented by the presence of foreground contamination: in particular, thermal dust and synchrotron emission from the Galaxy dominate over the primordial B-modes at high and low frequencies, respectively (see, e.g., [4]). Several experiments have been planned to detect B-mode polarization, either from ground, balloons or space. In this work, we focus on the forthcoming LSPE mission, that will be devoted to the accurate measurement of CMB polarization at large angular scales in order to constrain r down to 0.03 [5]. LSPE consists of two instruments: SWIPE, a balloon-borne array of bolometric polarimeters that will map the sky at 140, 220 and 240 GHz, and STRIP, a ground-based array of coherent detectors that will survey the same sky region at 43 and 90 GHz. This work focuses on the SWIPE experiment [6]. SWIPE will be likely launched during the winter 2018-2019 from the Svalbard islands and will operate for around 15 days during the Arctic night. A large fraction of the northern sky (around 30%) will be observed with angular resolution of about 1.5 deg Full Width at Half Maximum (FWHM). The 140 GHz band will be the main CMB science channel, while measurements at 220 and 240 GHz will be devoted to monitoring thermal dust contamination. The goal of this work is to provide a preliminary, but realistic, estimate of the foreground impact on LSPE-SWIPE observations.

2. Methodology and results

2.1. From Planck to LSPE-SWIPE

In order to infer the level of foreground contamination in the sky patch and at the frequency channels of interest for LSPE-SWIPE, we use the public foreground maps from the Planck satellite as templates [7]. In particular, we take the thermal dust and synchrotron maps estimated by the component separation method Commander [8] at reference frequencies of 353 and 30 GHz, respectively (see Fig. 1).

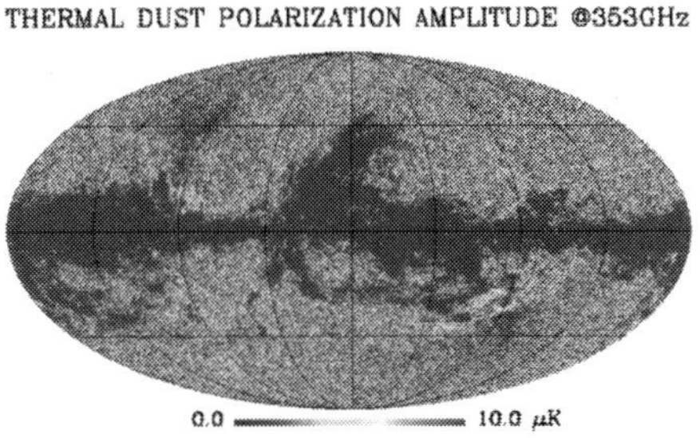

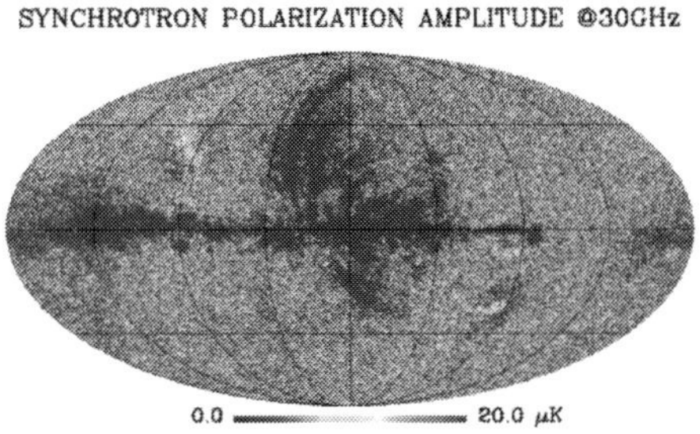

Figure 1. Thermal dust (left) and synchrotron (right) polarization amplitude maps ($P = \sqrt{Q^2 + U^2}$, where Q and U are the Stokes parameters for the linear polarization) from Planck. Maps are estimated by the component separation method Commander at reference frequencies of 353 and 30 GHz, respectively. Maps are in Galactic coordinates, antenna temperature units (μK_{RJ}) and a resolution of 10 arcmin for dust and 40 arcmin for synchrotron.

To first order, the dust spectrum is a modified Black Body (Gray Body) of the shape $\nu^{\beta_d} B_\nu$, where ν is the frequency, β_d is the dust spectral index and B_ν is the Planck function. Synchrotron emission is instead characterized by a power law spectrum with spectral index β_s. Hence, we rescale the dust and synchrotron maps to the LSPE-SWIPE frequency channels ($\nu = 140, 220, 240$ GHz) multiplying them by the following factors, respectively:

$$\alpha_\nu^d = \left(\frac{\nu}{\nu_{353\text{GHz}}}\right)^{\beta_d - 2} \frac{B_\nu(T_d)}{B_{\nu_{353\text{GHz}}}(T_d)}, \qquad \alpha_\nu^s = \left(\frac{\nu}{\nu_{30\text{GHz}}}\right)^{\beta_s} \tag{1}$$

where the dust and synchrotron spectral indices are $\beta_d = 1.58$ and $\beta_s = -3.04$ and the dust temperature is $T_d = 19.6$ K [4]. We convert the maps from the original antenna temperature to the more common thermodynamic temperature multiplying by the factor $g = \frac{(\exp(x)-1)^2}{x^2\exp(x)}$ where $x = \frac{h\nu}{kT_{CMB}}$, with h, k and T_{CMB} being the Boltzmann and Planck constants and the CMB average temperature, respectively. Finally, using the HEALPix package [9], we smooth and degrade the maps to 1.5 deg FWHM and to a map resolution parameter of $N_{side} = 128$, where N_{side} defines the total number of pixels N_{pix} in the map according to $N_{pix} = 12N_{side}^2$. In Fig. 2, we show realistic simulated maps of the CMB polarization amplitude, instrumental noise and foreground in the sky patch which will be observed by LSPE-SWIPE at 140 GHz. The CMB map has been produced with the HEALPix Synfast facility from theoretical power spectra calculated with the CAMB software [10] according to the Planck latest release of cosmological parameters [1] and assuming a tensor-to-scalar ratio $r = 0.03$. The noise map has been produced from a noise spectrum with a high frequency constant plateau of amplitude 15 μKs$^{1/2}$, and a low-frequency correlated $1/f^2$ part. Moreover, we include a low-frequency $1/f^2$ cross-correlated noise component equally shared by all the detectors. The knee frequency is set to $f_k = 0.1$ Hz. The noise map has been estimated with the ROMA map-making code [11] according to the nominal instrumental baseline and scanning strategy. The foreground (dust+synchrotron) map is produced according to the procedure described above.

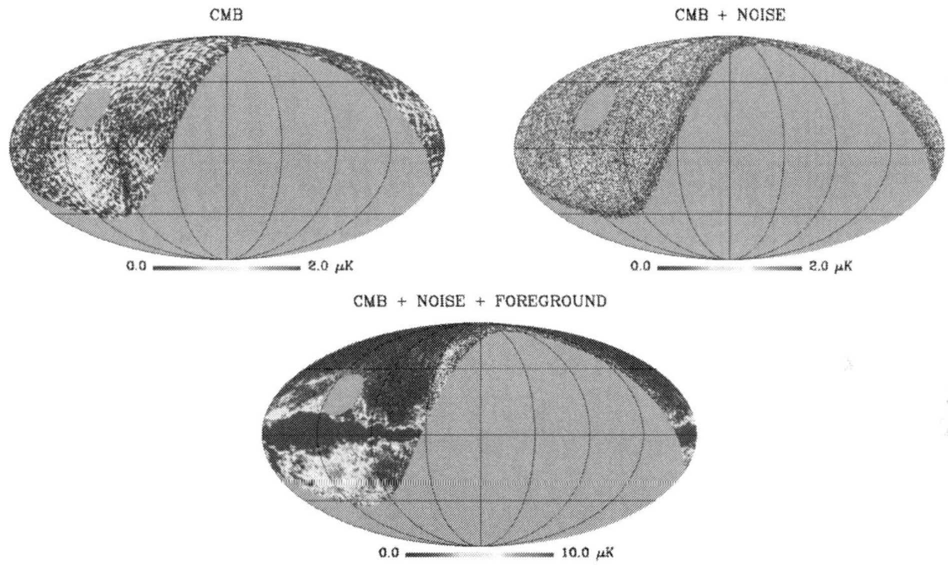

Figure 2. Simulated CMB (top left), CMB+noise (top right) and CMB+noise+foreground (bottom) polarization amplitude maps evaluated in the LSPE-SWIPE sky region at 140 GHz. Maps are in Galactic coordinates, thermodynamic temperature units (μK) and a resolution of 1.5 deg.

2.2. Forecasts on foreground contamination
In order to provide a quantitative assessment of the impact of foregrounds on LSPE-SWIPE observations, we estimate power spectra of the foreground maps generated as described in the previous section, but keeping the maps at native resolutions. In this process, masking the Galactic plane is a crucial step, as the foreground emission in that region is several orders of magnitude higher than the CMB signal we are interested in (as evident in Fig. 2). Galactic and extra-galactic polarized point sources need also to be masked out, as despite their low number they can alter the evaluation of the diffuse foreground power spectra. We use point source masks constructed from the polarized sources at 30 and 353 GHz in the Planck PCCS2 catalogue [12],

for the synchrotron and dust templates, respectively. It is worth mentioning that the Planck maps we use as templates do actually contain noise. Therefore, to avoid any bias in our analysis, we compute the spectra by cross-correlating maps generated from the Half-Mission (HM) data splits, which have independent noise. In particular, we found that the effect of point sources and noise is subdominant for the dust maps in the multipole range considered here ($2 \leq \ell \leq 100$), while synchrotron maps are significantly affected by both. Spectra in this paper have been estimated using the MASTER approach [13, 14]. In order to validate the methodology for our analysis setup, we simulated a set of foreground maps as random Gaussian realisations of an input power spectrum of the form $C_\ell = A^{BB}(\ell/\ell_0)^{-2.4}$ with $A^{BB} = 0.5A^{EE}$ [15]. We estimated the power spectra of these maps on the masks used in the paper and verified that the mean recovered power spectra are actually unbiased. In Fig. 3, we show the dust and synchrotron BB spectra estimated in the LSPE-SWIPE sky region at 140, 220 and 240 GHz, produced either without or including a temperature Galactic mask (the Planck GAL70 mask), which reduces the effective sky fraction from 36% to 24%. As expected, the use of optimized Galactic masks and component separation methods will be unavoidable to face foreground contamination.

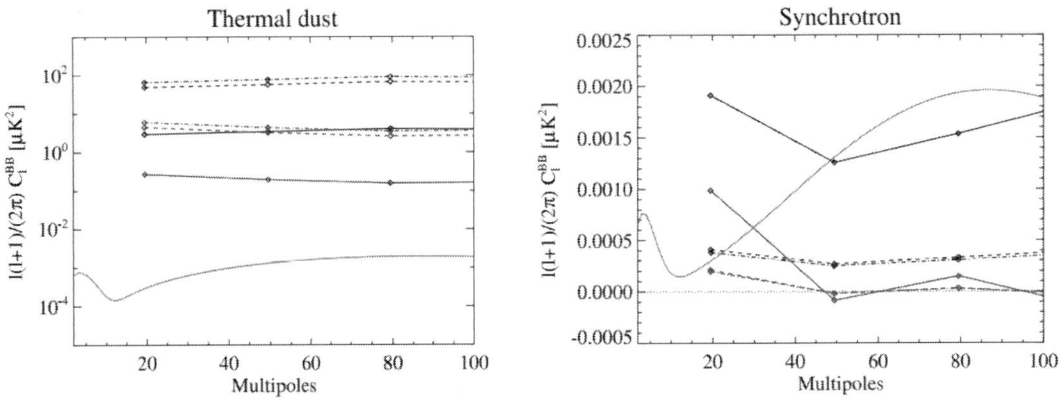

Figure 3. Thermal dust (left) and synchrotron (right) BB angular power spectra estimated in the LSPE-SWIPE sky region at 140 (solid line), 220 (dashed line) and 240 (dashed-dotted line) GHz, including (in blue) or not (in black) a temperature Galactic mask (the Planck GAL70 mask). We compute the spectra in three multipole bins for the ranges 5-34, 35-64 and 65-94. The theoretical CMB spectrum corresponding to $r = 0.03$ is shown in red for comparison.

To refine this analysis, we generate a set of polarization specific Galactic masks by thresholding a map obtained summing in quadrature the polarization intensity of the two foreground components at 140 GHz, built as $\sqrt{P^2_{dust} + P^2_{synch}} = \sqrt{Q^2_{dust} + U^2_{dust} + Q^2_{synch} + U^2_{synch}}$, after smoothing the original maps of the Stokes parameters to an effective resolution of 1.5 deg. We mask all the pixels in the map that are above certain thresholds, which we choose to be 0.75, 1, 1.5, 2, 2.5 and 3 σ_{CMB}, where $\sigma_{CMB} = \sqrt{\sigma^2(Q) + \sigma^2(U)} \simeq 0.38$ μK has been estimated from a simulation with $r = 0.03$ and the same resolution of the data maps. Finally, we smooth the masks with a Gaussian beam of 10 deg FWHM, put a threshold at 0.5 to have masks with only [0,1] values and combine these masks with the LSPE-SWIPE targeted sky patch. In Fig. 4, we show the six sky regions selected by these masks, which have effective sky coverages ranging from 1.3 to 15.1%. The choice of a proper Galactic mask must be done according to the specific requirements on the sky fraction and the residual foreground level, which will depend on the component separation method. Clearly, the effective sky fraction should be maximised to reduce cosmic variance, while the foreground amplitude should be mitigated as far as possible.

We then estimate the dust and synchrotron BB power spectra at 140 GHz for the different masks and we assess the uncertainties due to noise in Planck data from Full Focal Plane simulations [16]. We consider a set of 100 HM simulated noise maps both at 30 and 353 GHz and extrapolate them in frequency as it has been done for the data (see Eq. 1). We notice that these simulations underestimate the variance of the data by a (scale-dependent) factor of 5-20% [16]. The spectra have been computed in three multipole bins for the ranges 5-34, 35-64 and 65-94 and are shown in Fig. 5.

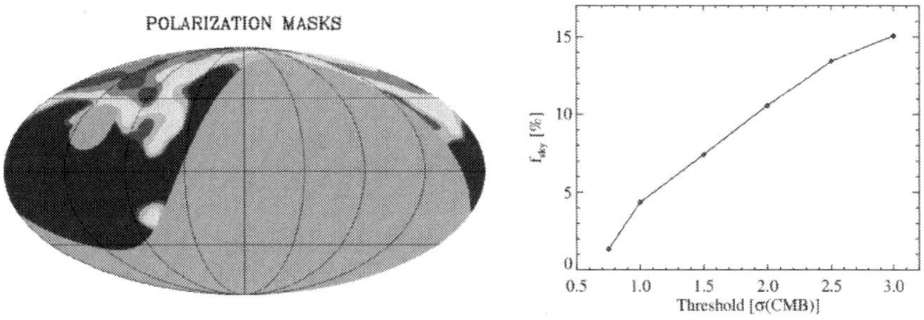

Figure 4. On the left: Masks obtained combining different masks of the Galactic plane with the LSPE-SWIPE scanning strategy. The Galactic masks are optimized in polarization assuming a frequency of 140 GHz and correspond to different thresholds: 0.75, 1, 1.5, 2, 2.5 and 3 σ_{CMB}. On the right: Plot of the sky fraction as a function of the threshold. See text for details.

Finally we calculate an equivalent tensor-to-scalar ratio for the foreground emission by dividing these spectra by the theoretical CMB BB spectrum with $r = 1$. In particular, we average through inverse-variance weighting the values of the equivalent tensor-to-scalar ratios from the two multipole bins ranging from $35 < \ell < 64$ and $65 < \ell < 94$ around the "recombination" bump, while we discard the lowest multipole bin as the Planck polarization data of the 2015 release still contain some low level residual systematics on those angular scales. Actually, it will be interesting to extend the analysis once the final Planck data, optimized for polarization on large scales, will be released at the end of 2017. Results are shown in Fig. 6 for the different sky fractions.

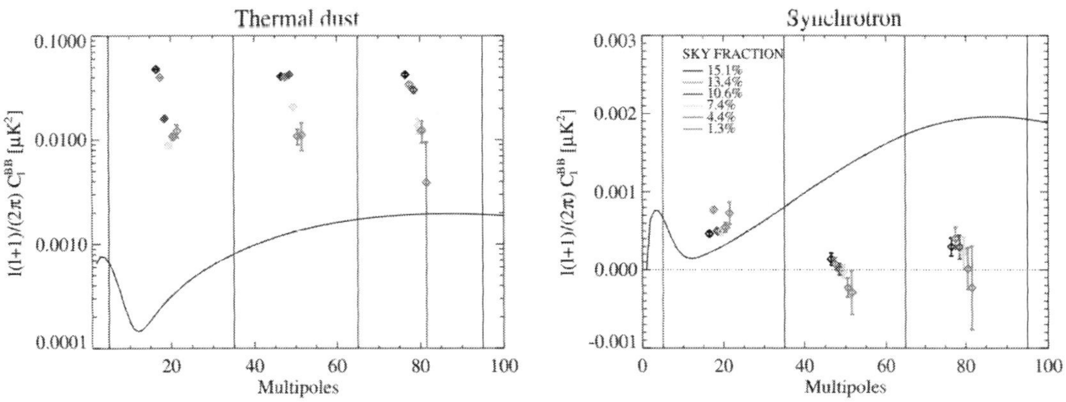

Figure 5. Thermal dust (left) and synchrotron (right) BB angular power spectra estimated in the LSPE-SWIPE sky region at 140 GHz, applying the Galactic masks given in Fig. 4. We compute the spectra in three multipole bins for the ranges 5-34, 35-64 and 65-94. The theoretical CMB spectrum corresponding to $r = 0.03$ is shown for comparison with a solid black line.

3. Conclusions

A CMB B-mode polarization detection would be a "smoking gun" evidence for inflation theory, as primordial B-modes are unambiguosly sensitive to the tensor mode of inflationary perturbations.

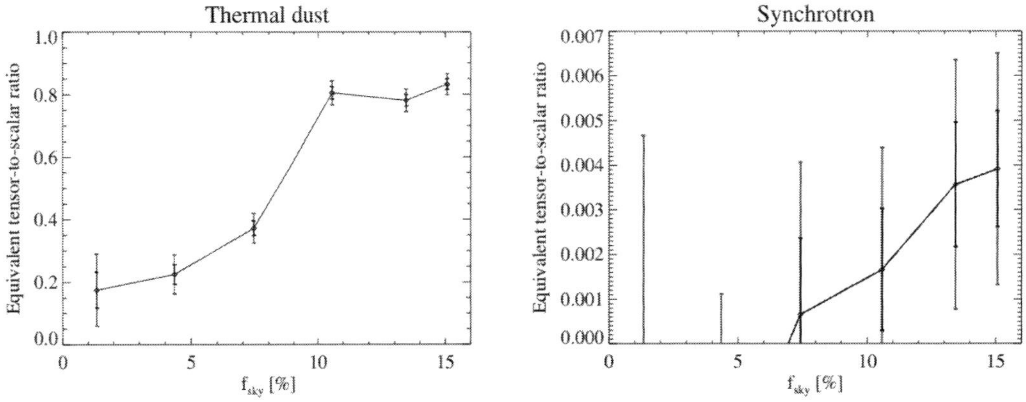

Figure 6. Equivalent tensor-to-scalar ratios vs sky fraction at 140 GHz for dust (left) and synchrotron emission (right). Error bars correspond to 1σ (in black) and 2σ (in blue) noise uncertainties.

A major obstacle to observe B-modes is represented by Galactic contamination of polarized thermal dust and synchrotron emission. In this work, using the latest Planck polarization data, we present preliminary forecasts on the impact of foregrounds on the observations of the upcoming LSPE-SWIPE balloon mission. Focusing on the main CMB channel (140 GHz) we find that, in the LSPE-SWIPE sky patch, the dust emission has a power spectrum larger than the target B-mode spectrum ($r = 0.03$) by around ~ 3 orders of magnitude across the entire multipole range we consider. Synchrotron emission is instead above the target B-modes only at the lowest multipoles, while it becomes sub-dominant on smaller scales. We generate polarization specific Galactic masks and verify that when preserving large sky fractions, at least 15%, the level of dust contamination is about a factor of 30 larger than the target primordial CMB B-mode signal. Therefore, our analysis shows that a foreground cleaning method able to reduce the foregrounds at the level of few percent will be required in order to disantangle the primordial signal from the Galactic contamination, and hence maximise the scientific return of the LSPE-SWIPE mission.

Acknowledgments

The LSPE project is supported by the contract I/022/11/0 of the Italian Space Agency and by INFN CSN2. We acknowledge the CINECA award under the ISCRA initiative, for the availability of high performance computing resources and support.

References

[1] Ade P A R *et al* [Planck collaboration] 2016 *A&A* **594** A13
[2] Ade P A R *et al* [BICEP2/Keck and Planck Collaborations] 2015 *Phys. Rev. Lett.* **114** 101301
[3] Kamionkowski M and Kovetz E D 2015 arXiv:1510.06042
[4] Ade P A R *et al* [Planck collaboration] 2015 *A&A* **576** A107
[5] Aiola S *et al* 2012 *SPIE Conf. Series* **8446** 277 arXiv:1208.0281v1
[6] de Bernardis P *et al* 2012 *SPIE Conf. Series* **8452** 125 arXiv:1208.0282v1
[7] http://pla.esac.esa.int/pla/#maps
[8] Adam R *et al* [Planck collaboration] 2016 *A&A* **594** A10
[9] Gorski K M *et al* 2005 *Astrophys. J.* **622** 759

[10] Lewis A and Bridle S 2002 *Phys. Rev. D* **66** 103511
[11] de Gasperis G *et al* 2016 *A&A* **593** A15
[12] Ade P A R *et al* [Planck collaboration] 2016 *A&A* **594** A26
[13] Hivon E *et al* 2002 *Astrophys. J.* **567** 2
[14] Polenta G *et al* 2005 *JCAP* **11** 001
[15] Adam R *et al* [Planck collaboration] 2016 *A&A* **586** A133
[16] Ade P A R *et al* [Planck collaboration] 2016 *A&A* **594** A12

Cosmology with 21cm intensity mapping

I P Carucci [1,2,3]

[1] SISSA - International School for Advanced Studies, Via Bonomea 265, 34136 Trieste, Italy
[2] INFN sez. Trieste, Via Valerio 2, 34127, Trieste, Italy
[3] LUTH, Observatoire de Meudon, 5 Place Janssen, 92195 Meudon, France

E-mail: ipcarucci@sissa.it

Abstract. The nature of the most abundant components of the Universe, dark energy and dark matter, is still to be uncovered. I tackle this subject considering a novel cosmological probe: the neutral hydrogen emitted 21cm radiation, observed with the intensity mapping technique. I analyse competitive and realistic dark energy and dark matter models and show how they produce distinctive and detectable effects on the 21cm signal. Moreover, I provide radio telescope forecasts showing how these models will be distinguishable in an unprecedented way.

1. Introduction

The current standard cosmological model, ΛCDM, has outstandingly provided the theoretical framework to interpret a wide variety of observables [1], however this success is linked to the existence of two dark components. The Universe is now experiencing an accelerated expansion generated by an unknown Dark Energy (DE) component [2, 3], while an invisible Dark Matter (DM) drives the formation of all visible structures (see e.g. [4–6]). Effects of both components are embedded in the spatial distribution of matter in the Universe, that we can trace thanks to different observables such as galaxy clustering, weak lensing, abundance of galaxy clusters and Lyman-α forest.

In the upcoming years, observations of the redshifted 21cm radiation from neutral hydrogen (HI) will allow us to map the large scale structure of the Universe with an unprecedented precision on redshifts not accessible with the above probes [7, 8]. The idea is to carry out Intensity Mapping (IM) observations: measuring with low angular resolution the integrated 21cm emission coming from unresolved sources contained in large patches of the sky. Given its spectroscopic nature and the large volumes sampled, with 21cm IM we will soon start doing cosmology. Among current and upcoming instruments able to employ this technique we find: CHIME[1], BINGO[2], ORT[3], FAST[4] and ultimately SKA[5] and its pathfinders.

In this paper, I will illustrate the imprint on the 21cm IM signal power spectrum left by non-standard cosmological models otherwise indistinguishable from ΛCDM, quantifying these effects in the context of realistic radio-telescope forecasts. The work presented is based on [9, 10].

[1] http://chime.phas.ubc.ca/
[2] http://www.jb.man.ac.uk/research/BINGO/
[3] http://rac.ncra.tifr.res.in/
[4] http://fast.bao.ac.cn/en/
[5] https://www.skatelescope.org/

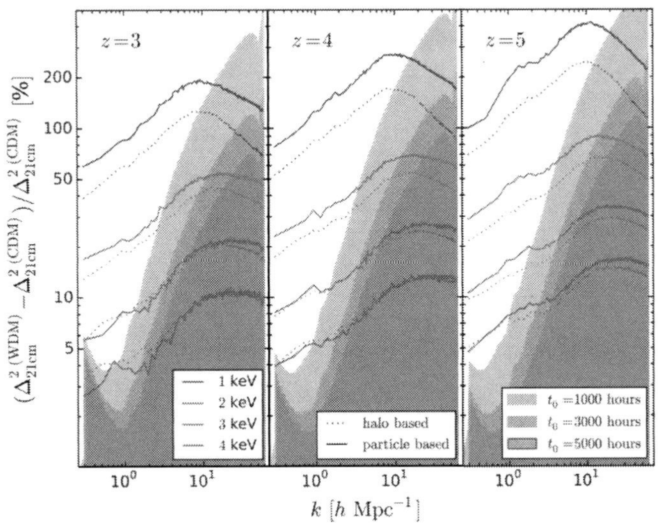

Figure 1. 21cm intensity map power spectrum of the thermal WDM models relative to the reference ΛCDM predictions at $z = 3$ (left), $z = 4$ (middle) and $z = 5$ (right), dotted and solid lines refer to different HI distribution methods (see [9] for details). The shaded area represents the expected errors from SKA1-LOW measurements for the reference ΛCDM model, $\sigma[P^{\Lambda CDM}_{21cm}(k)]/P^{\Lambda CDM}_{21cm}(k)$, assuming $t_0 = 1000$ (grey), $t_0 = 3000$ (blue) and $t_0 = 5000$ (fuchsia) observing hours.

2. Modelling the HI spatial distribution

Radio telescopes detect the redshifted 21cm radiation emitted by neutral hydrogen, thus we need to model the distribution of HI in the Universe. Here I summarise how we carry out the modelling and the analysis. After cosmic reionization, it is believed that the quasi totality of HI resides in DM halos, that are most accurately modelled with numerical simulations being halos biased tracers of the DM density field. We make use of simulations that we present together with the cosmologies considered. After having identified the halos (in all cases with a FoF algorithm [11]), we proceed at populating them with HI following the prescriptions presented in [9, 10]. Once the gas is distributed, computing the expected 21cm signal is straightforward. To better assess the significance of the computed signals, we forecast errors with which SKA1-LOW and SKA1-MID[6] will measure them, taking care of the instrumental noise only, i.e. assuming that astrophysical foregrounds, radio frequency interference and others nuisances have been already removed from the observed data.

3. Non-standard dark matter models

The ΛCDM model is constituted by a perfectly collisionless *cold* dark matter component (CDM), i.e. with negligible thermal velocities on all scales at high redshift. However, tensions exist at small scales between CDM predictions and various observational probes (see [12] for a review). There are various theoretical DM models that can alleviate these tensions while preserving the success of CDM on large scales, i.e. they are indistinguishable from CDM at large scales. They do so by suppressing gravitational clustering at small scales, i.e. DM can not significantly cluster below a specific length. The overall effect is to retain a matter power spectrum with a cut-off on a characteristic scale that depends on the nature of the DM particle. Here we take into consideration three families of models:

(i) WDM: thermal warm DM, with intrinsic momenta derived from a Fermi-Dirac distribution;
(ii) ULADM: ultra-light axion DM models, in the literature sometime called *fuzzy* DM [13];
(iii) LFDM: late-forming DM models [14, 15].

All these models are characterised by the above invoked suppression of the amplitude of matter density fluctuations at small scales below a characteristic length that for WDM models

[6] The choice of the telescope depends on the considered redshift of observation.

IOP Conf. Series: Journal of Physics: Conf. Series **956** (2017) 012003 doi:10.1088/1742-6596/956/1/012003

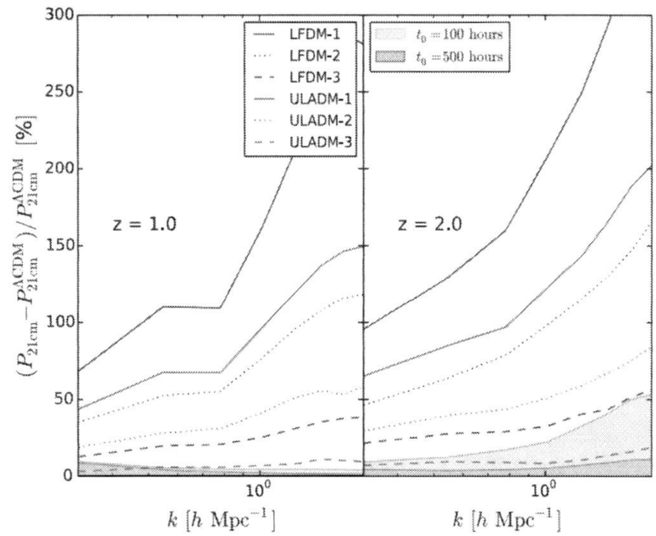

Figure 2. 21cm intensity map power spectrum of the LFDM and ULADM models relative to the reference ΛCDM-S predictions at $z = 1$ (left panel) and 2 (right panel) respectively. The shaded area represents the expected errors from SKA1-MID measurements for the reference ΛCDM-S model, $\sigma[P_{21cm}^{\Lambda CDM}(k)]/P_{21cm}^{\Lambda CDM}(k)$, assuming $t_0 = 100$ (light shaded area) and $t_0 = 500$ (dark shaded area) observing hours. Figure taken from [10].

depends on the thermal relic mass, for ULADM on the axion particle mass, while in the case of LFDM models depends on the phase transition redshift.

In the case of WDM, we consider a set of five high-resolution hydrodynamic N-body simulations presented in [9]. They follow the evolution of 512^3 CDM/WDM and 512^3 baryon (gas+stars) particles within simulation boxes of comoving sizes equal to 30 h^{-1}Mpc. We have simulated five different cosmological models: one model with CDM and four models with WDM, each of them with a different Fermi-Dirac mass of the WDM particles: 1 keV, 2 keV, 3 keV and 4 keV. This keV range is that of interest since today tightest constraints come from the Lyman-α forest high redshift power spectrum and point towards lower limits of $m_{\rm WDM} \gtrsim 3.5 - 5.3$ keV (at 2σ C.L.) [16]. The values of the cosmological parameters are common to all five simulations: $\Omega_{\rm m} = 0.3175$, $\Omega_{\rm b} = 0.049$, $\Omega_\Lambda = 0.6825$, $h = 0.6711$, $n_s = 0.9624$ and $\sigma_8 = 0.834$. We analyse snapshots at redshifts $z = 1$, 2 and 3.

As shown in [9], for WDM masses between 3 and 4 keV we observe a suppression in power up to $\sim$ 10% on very small scales and a reduction in the number of 10^9 h^{-1}M$_\odot$ halos of the order of 20-40% compared to the CDM case. This lack of low mass halos, results in a higher amplitude of the 21cm power spectrum for the WDM than CDM, because the spatial distribution of neutral hydrogen is more strongly clustered in the models with WDM. In Fig. 1 we show the relative difference in the signal power spectrum between the models with WDM and the model with CDM. The uncertainties with which the SKA1-LOW radio-telescope will measure the 21cm power spectrum are highlighted with shaded regions. We find that with a reasonable observational time of $t_0 = 1000$ hours the WDM models with 1, 2, and 3 keV can be distinguished from CDM. On the other hand, the predicted power spectrum of the 4 keV WDM model is consistent with the one predicted by CDM at $\sim 1\sigma$ confidence level at $z = 5$.

For the last two families of non-standard DM, ultra-light axions and late forming models, we consider N-body simulations presented in [17] with box of comoving size of 27.5 Mpc h^{-1}, that follow 1024^3 particles. They consist of three ULADM models[7] : with axion mass of $m_a = 1.56 \times 10^{-22}$ eV (ULADM-1), 4.16×10^{-22} eV (ULADM-2) and 1.54×10^{-21} eV (ULADM-3), and of three late-forming DM models with transition redshift $z_t = 5 \times 10^5$ (LFDM-1), 8×10^5 (LFDM-2) and 15×10^{15} (LFDM-3). The cosmological model parameters have been set to those

[7] Note that the ULADM models investigated are in disagreement with the recent constraints obtained at $z > 3$ from the Lyman-α forest which result in a lower limit at the 2σ level of $\sim 2 \times 10^{-21}$ eV [18].

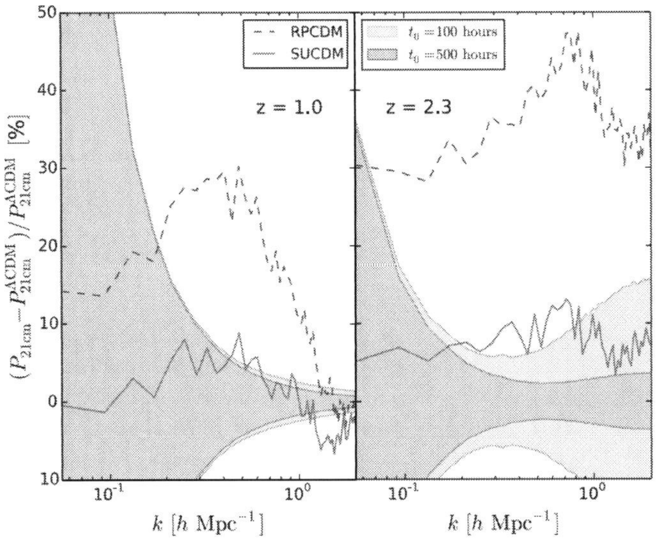

Figure 3. As in Fig. 2 for the non-standard DE models relative to ΛCDM-W5 model. Figure taken from [10].

of a reference ΛCDM simulation (ΛCDM-S) of the same box size and with equal number of particles with $\Omega_m = 0.3$, $\Omega_b = 0.046$, $h = 0.7$, $\sigma_8 = 0.8$ and $n_s = 0.99$. We analyse snapshots at $z = 1$ and 2.

As for WDM, also these models are characterised by a suppression in the number of low mass halos. In Fig. 2 we plot the 21cm IM spectra in the case of the ULADM and LFDM models relative to the reference ΛCDM-S. Again, we address significance by showing the forecasted error for the SKA1-MID radio telescope in interferometry for the reference ΛCDM-S model 21cm IM measurement.

The 21cm power spectra of the non-standard DM cosmologies analysed do not display any small scale cut-off as the matter power spectra do and this is consequence of neutral hydrogen being hosted in DM halos: these scenarios suppress halos abundance and force HI to be more clustered in the halos that are left (that are also the more massive, i.e. more biased). This results in expecting a higher 21cm IM signal in these scenarios, enough higher than instrumental noise to tell them apart from a standard CDM paradigm (as shown in Figs. 1-2).

4. Dynamical dark energy

We use a set of large volume N-body simulations from the Dark Energy Universe Simulations (DEUS) database[8] of three flat DE models:

(i) ΛCDM-W5: a standard cosmological model with cosmological constant Λ,
(ii) RPCDM-W5: a quintessence model with dynamical equation of state as given by the scalar field evolution in a Ratra-Peebles [19] self-interacting potential,
(iii) SUCDM-W5: a quintessence model with supergravity [20] self-interacting potential.

The cosmological parameters of all above models have been calibrated in [21] in order to reproduce within 1σ the cosmic microwave background power spectra from WMAP-5 observations [22] and the luminosity distances from SN-Ia measurements [23]. In Table 1 we show the tuned cosmological parameters (with w_0 and w_a being the DE equation of state parameters of the Linder-Chevalier-Polarski parametrization [24, 25] that best-fit the time evolution of the quintessence-field equation of state); the other cosmological parameters are set to $\Omega_b = 0.04$,

[8] http://www.deus-consortium.org/deus-data/

IOP Conf. Series: Journal of Physics: Conf. Series **956** (2017) 012003 doi:10.1088/1742-6596/956/1/012003

$h = 0.72$, $n_s = 0.96$ for all models considered. All simulations have comoving box size of 162 Mpc h^{-1} and follow the evolution of 1024^3 particles. We analyse snapshots at $z = 1$ and 2.3.

Model	Ω_m	σ_8	w_0	w_a
ΛCDM-W5	0.26	0.80	-1	0
RPCDM-W5	0.23	0.66	-0.87	0.08
SUCDM-W5	0.25	0.73	-0.94	0.19

Table 1. Cosmological model parameters of realistic DE models calibrated against WMAP-5 and SN Ia observations. Details in [21].

As discussed above, the quintessence models investigated are indistinguishable from ΛCDM using the considered cosmological data sets. They are characterized by small differences in the large scale clustering that get amplified at small scales as the non-linear regime of gravitational collapse begins [21]. Being quantitative, the quintessence models exhibit DM density power spectra in the range $0.1 \lesssim k \lesssim 1\,\mathrm{Mpc}\,h^{-1}$ and $z \lesssim 2$ of $20 - 40\%$ lower amplitude than the ΛCDM prediction. This is caused by the onset of DE dynamics that alters the cosmic expansion during the late matter dominated era by causing a more accelerated expansion at this epoch than in the standard ΛCDM case. Therefore matter density fluctuations grow less efficiently and the dynamical DE models exhibit a lower level of clustering and, with direct consequences for the HI distribution, lower halo abundances: on average the RPCDM-W5 model has larger deviations ($\sim 20 - 80\%$) than SUCDM-W5 ($\sim 5 - 40\%$), as reported in [10], consistently with the linear growth rate of these models.

In Fig. 3 we plot the relative difference of the 21cm power spectrum between the non-standard DE models and the reference ΛCDM-W5 at $z = 1$ (left panel) and $z = 2.3$ (right panel). The RPCDM model can be distinguished from the ΛCDM at high statistical significance. Even the SUCDM model, that features a cosmic expansion and a linear growth rate similar to that of the ΛCDM, can be potentially distinguished at more that 1σ at $z = 2.3$ in the range of scales corresponding to $0.02 \lesssim k \lesssim 2\,\mathrm{Mpc}^{-1}\,h$.

Looking together at Figs. 1-2-3, we notice that: *i)* the 21cm spectra of the DE models differ from the ΛCDM case not only in amplitude but also in the scale dependence of the signal, especially at small scales where the slope changes; *ii)* the non-standard DM models differ from the standard cosmological scenario following another trend, with differences increasing at small scales. This suggests that in principle imprints of DE and DM models can be distinguished from one another through 21cm IM observations.

5. Conclusions

In this paper I presented some features of the expected 21cm radiation signal in intensity mapping for cosmologies with non-cold DM and non-Λ DE.

I quantified the constraints that actual radio-surveys will be able to set on such models, that are otherwise statistically indistinguishable from standard ΛCDM with current observational evidences (this holds at all scales for the dynamical DE cases and at large scales for the non-standard DM cases).

The model dependent features of each scenario are manifest in the 21cm IM spectra such that future SKA measurements should be able to detect or rule out the non-standard DE and DM models we have considered.

Acknowledgments

I am happy to thank my collaborators: Matteo Viel, Francisco Villaescusa-Navarro, Pier-Stefano Corasaniti and Andrea Lapi. I also want to thank the students of the Physics Department of the University of Cagliari, who have been amazing hosts of this meeting.

References

[1] Planck Collaboration 2016 *Astronomy and Astrophysics* **594** A13 (*Preprint* 1502.01589)
[2] Riess A G *et al* 1998 *The Astronomical Journal* **116** 1009–1038 (*Preprint* astro-ph/9805201)
[3] Perlmutter S *et al* 1999 *Astrophysical Journal* **517** 565–586 (*Preprint* astro-ph/9812133)
[4] Tegmark M *et al* 2004 *Physical Review D* **69** 103501 (*Preprint* astro-ph/0310723)
[5] Clowe D, Bradač M, Gonzalez A H, Markevitch M, Randall S W, Jones C and Zaritsky D 2006 *Astrophysical Journal Letters* **648** L109–L113 (*Preprint* astro-ph/0608407)
[6] Massey R *et al* 2007 *Nature* **445** 286–290 (*Preprint* astro-ph/0701594)
[7] Bull P, Ferreira P G, Patel P and Santos M G 2014 *ArXiv e-prints* (*Preprint* 1405.1452)
[8] Camera S, Santos M G, Ferreira P G and Ferramacho L 2013 *Physical Review Letters* **111** 171302 (*Preprint* 1305.6928)
[9] Carucci I P, Villaescusa-Navarro F, Viel M and Lapi A 2015 *Journal of Cosmology and Astroparticle Physics* **7** 047 (*Preprint* 1502.06961)
[10] Carucci I P, Corasaniti P S and Viel M 2017 *ArXiv e-prints* (*Preprint* 1706.09462)
[11] Davis M, Efstathiou G, Frenk C S and White S D M 1985 *Astrophysical Journal* **292** 371–394
[12] Weinberg D H, Bullock J S, Governato F, Kuzio de Naray R and Peter A H G 2015 *Proceedings of the National Academy of Science* **112** 12249–12255 (*Preprint* 1306.0913)
[13] Marsh D J E 2016 *Physics Reports* **643** 1–79 (*Preprint* 1510.07633)
[14] Das S and Weiner N 2011 *Physical Review D* **84** 123511 (*Preprint* astro-ph/0611353)
[15] Agarwal S, Corasaniti P S, Das S and Rasera Y 2015 *Physical Review D* **92** 063502 (*Preprint* 1412.1103)
[16] Iršič V *et al* 2017 *ArXiv e-prints* (*Preprint* 1702.01764)
[17] Corasaniti P S, Agarwal S, Marsh D J E and Das S 2017 *Physical Review D* **95** 083512 (*Preprint* 1611.05892)
[18] Iršič V, Viel M, Haehnelt M G, Bolton J S and Becker G D 2017 *ArXiv e-prints* (*Preprint* 1703.04683)
[19] Ratra B and Peebles P J E 1988 *Physical Review D* **37** 3406–3427
[20] Brax P and Martin J 2000 *Physical Review D* **61** 103502 (*Preprint* astro-ph/9912046)
[21] Alimi J M, Füzfa A, Boucher V, Rasera Y, Courtin J and Corasaniti P S 2010 *Montly Notices of the Royal Astronomical Society* **401** 775–790 (*Preprint* 0903.5490)
[22] Komatsu E *et al* 2009 *Astrophysical Journal Supplement* **180** 330–376 (*Preprint* 0803.0547)
[23] Kowalski M *et al* 2008 *Astrophysical Journal* **686** 749-778 (*Preprint* 0804.4142)
[24] Linder E V 2003 *Physical Review Letters* **90** 091301 (*Preprint* astro-ph/0208512)
[25] Chevallier M and Polarski D 2001 *International Journal of Modern Physics D* **10** 213–223 (*Preprint* gr-qc/0009008)

Study of the eclipse region of the redback millisecond pulsar J1431−4715

A Miraval Zanon[1], M Burgay[2], A Possenti[2] and A Ridolfi[3]

[1] Dipartimento di Fisica e Astronomia "Galileo Galilei" - Università di Padova, Vicolo dell'Osservatorio 3, Padova, IT-35122, Italy
[2] Osservatorio Astronomico di Cagliari, INAF, Via della Scienza 5, I-09047 Selargius (CA), Italy
[3] Max-Planck-Institut für Radioastronomie, Auf dem Hügel 69, D-53121 Bonn, Germany

E-mail: arianna.miravalzanon@studenti.unipd.it

Abstract. We report on the rotational, astrometric and orbital parameters for PSR J1431−4715, and we also present a preliminary analysis of the eclipsing region. This pulsar was discovered in the High Time Resolution Universe survey and it belongs to the class of "redback" systems. The minimum estimated mass for the companion of J1431−4715 is, indeed, 0.13 $M_\odot$. Thanks to multi-frequency observations, obtained at the 64 m Parkes radio telescope, we note that the magnitude and the duration of the eclipse delay depend upon the observing frequency.

1. Introduction

MilliSecond pulsars (MSPs) are a population of Neutron Stars (NSs) with fast spin (hundreds of rotations per second) and relatively weak magnetic field (10^8 G).
These objects are thought to be generated in Low-Mass X-ray Binary systems (LMXBs), where a NS is spun up by the accretion of matter and angular momentum from its companion [1]. This process is called "recycling" and the pulsars that have experienced it are dubbed "recycled" pulsars. The X-ray emission of LMXBs is actually powered by this accretion.
According to the current theory, the progenitors of MSPs are bright X-ray binaries, in which a luminous main sequence star overflows its Roche lobe and transfers mass to the NS. Tidal forces during the mass transfer circularize the orbit, while the original companion star loses matter and typically becomes a white dwarf.
Among MSPs, the so-called "eclipsing binary millisecond pulsars" emit a radio signal that is distorted or completely disappears, at some particular orbital phases. There are two sub-populations of eclipsing MSPs: Black Widow (BW) and RedBack (RB) pulsars. BWs have very low-mass companions ($M_2 \leq 0.1$ $M_\odot$), while RBs have slightly heavier companions ($M_2 \simeq 0.1 - 0.4$ $M_\odot$). In the former type, the companion star is a largely ablated degenerate star; in the latter type, the companion is non-degenerate and only partially ablated.
The discovery of BW and RB systems has highlighted the link between LMXBs and recycled binary radio pulsars. Recent observations (e.g. [2]) confirmed the transition between an accretion-powered state (in which the system is a LMXB and bright X-ray emission is observed) and a rotation-powered state (in which radio emission is powered by the NSs rotating magnetic field) for some RB systems.

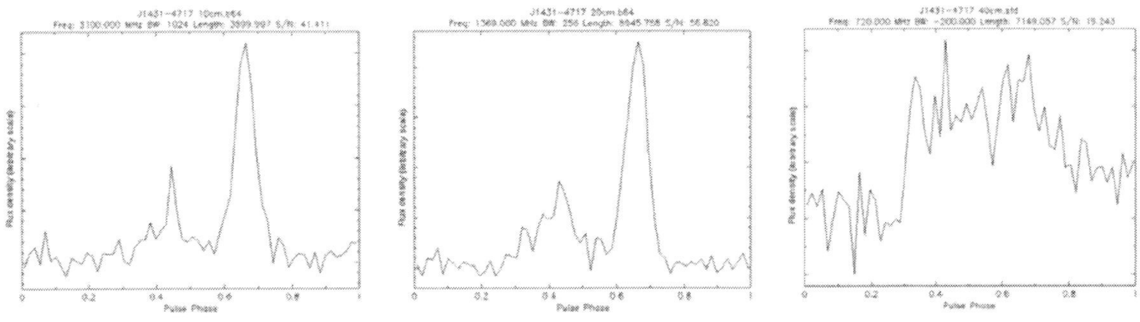

Figure 1. Starting from the left, integrated pulse profile of PSR J1431−4715 at 3.1 GHz, 1.4 GHz and 0.7 GHz.

Observations of the eclipse regions of BWs and RBs show that the radio pulsed signal is delayed, distorted or completely absorbed by a low-dense, highly ionized gas cloud enshrouding the companion. This eclipsing material can be originated by the companion star being hit by the pulsar's high-energy wind [3]. Alternatively, the eclipsing material could come from an accretion disk whose flow of matter is stopped and blown away by the MSP's radiation pressure at the inner Lagrangian point [4].

2. Observations and analysis

PSR J1431−4715 was discovered in the High Time Resolution Universe (HTRU) survey [5] with the 64 m Parkes radio telescope using the 13 beam 21 cm multibeam receiver, observing the intermediate Galactic latitude regions ($-120° < l < 30°$, $\mid b \mid \leq 15°$).
Our dataset consisted of observations made with the Parkes radio telescope from 2011 July to 2017 June. The pulsar was regularly observed in this time interval with different backends: APSR (Australia Telescope National Facility Parkes Swinburne Recorder), CASPSR (Center for Astronomy Signal Processing and Electronics Research Parkes Swinburne Recorder) and DFB3/4 (Digital Filterbank) [6]. Most of the observations were made at 1.4 GHz, while some others were made simultaneously at 0.7 GHz and 3.1 GHz with the coaxial 10/50 cm receiver. The pulsar is well visible at 1.4 GHz and 3.1 GHz, while the signal at 0.7 GHz is very noisy.
The pulsar signal, crossing the InterStellar Medium (ISM), suffers frequency-dependent delays due to the effects of dispersion (for more details see [7]). These delays can be corrected with a procedure called de-dispersion, which is based on the knowledge of the Dispersion Measure (DM). The DM represents the electron column density along the source line of sight and can be expressed as:

$$\mathrm{DM} = \int_0^d n_e(l)\mathrm{d}l, \tag{1}$$

where n_e is the electron density along the path, d is the source distance with respect to the Earth. Dispersion occurs because the group velocity of electromagnetic waves through an ionized medium is frequency dependent. Radiation at higher frequencies travels faster than radiation at lower frequencies.
In the observations that we analyzed the signal is acquired in *folding mode*. When folding, the observing band is first divided into a number of channels and in each one the effect of dispersion is removed according to the value of the DM. After de-dispersing, the observation is divided into subsequent time intervals, called sub-integrations. Each sub-integration is summed in phase according to the pulsar spin period, in order to form an integrated profile. This observing mode

is used when the basic characteristics of the pulsar are known, as in our case. The folding procedure allows us to sum many pulses in order to make the signal visible over the background noise.

The aim of our analysis was mainly to measure with high accuracy the astrometric, rotational and orbital parameters of PSR J1431−4715, using a procedure called *timing*.

Using the `PSRCHIVE` pulsar software [8], we have extracted the topocentric times-of-arrival (ToAs) by cross-correlating each integrated pulse profile with a high signal-to-noise (S/N) template profile. Template profiles at three different frequencies are shown in Fig. 1; these were obtained by coherently summing our best detections. Multi-frequency observations show a significant pulse profile evolution: pulse profiles at 3.1 GHz and 1.4 GHz are doubled peaked, while the pulse profile at 0.7 GHz is broadened and noisy.

We used the `TEMPO2` pulsar timing package [9] in order to convert the topocentric ToAs, t_i, to the barycentric ToAs $t_{i,bary}$, referred to the Solar System Barycenter (SSB), with the formula:

$$t_{i,bary} = t_i + t_{clock} - \frac{D(t_i)}{f^2} + \Delta R_{\odot} + \Delta E_{\odot} + \Delta S_{\odot}, \tag{2}$$

where t_{clock} is the clock correction chain, $D(t_i)$ is the delay due to interstellar dispersion, $\Delta R_{\odot}$ is the Roemer delay, $\Delta E_{\odot}$ is the Einstein delay and $\Delta S_{\odot}$ is the Shapiro delay.

In our case the pulsar is in a binary system, so before translating the ToAs expressed in pulsar proper time at the pulsar surface to the SSB, we must calculate them at the Pulsar Binary Barycenter.

The aim of the timing procedure is to find the parameters that correctly describe the rotation of our pulsar by accounting for every single rotation of the NS in a given time window. To do so, `TEMPO2` models the rotational evolution of the neutron star with a Taylor expansion of the number of rotations:

$$N(t) = \nu_{ep}(t - t_{ep}) + \frac{1}{2}\dot{\nu}_{ep}(t - t_{ep})^2 + \frac{1}{6}\ddot{\nu}_{ep}(t - t_{ep})^3 + \ldots, \tag{3}$$

where ν_{ep}, $\dot{\nu}_{ep}$, $\ddot{\nu}_{ep}$ are neutron star spin frequency, the first and second derivative of the spin frequency at a reference epoch t_{ep}. $N(t)$ represents the number of rotations occurred from t_{ep} to t.

Using again `TEMPO2` we examined the so-called *timing residuals*, i.e. the differences between the observed ToAs and the values predicted by our model. In order to derive the best timing solution we have minimized the weighted χ^2 for the astrometric, rotational and orbital parameters. The best-fitting timing parameters for J1431−4715 are reported in Tab. 1. Thanks to the large time span of our dataset, we were able to measure the orbital period derivative $\dot{P}_b$. We performed an F-test to determine if the addition of $\dot{P}_b$ was indeed required by the data. Without it, $\chi^2 = 440$ with 60 degrees of freedom, while after fitting for $\dot{P}_b$ the χ^2 improved to 71.12 with 70 degrees of freedom. A negative $\dot{P}_b$ indicates a gravitational energy loss by the system, but the contribution to $\dot{P}_b$ expected from General Relativity (GR) is three orders of magnitude smaller than observed, assuming $M_p = 1.4$ $M_{\odot}$ and $M_c = M_{c,\min}$. The value found for $\dot{P}_b$ with timing analysis is probably due to mass loss from the companion star. This lost matter is ejected from the system due to the pulsar wind and leads to a decrease of the angular momentum and the orbital period of the system.

The timing residuals shown in the left plot of Fig. 2, as a function of Modified Julian Day, (MJD), do not show any structure but are randomly distributed. This is indicative of the fact that the best-fitting timing model is correctly describing the pulsar's behaviour. The right plot in Fig. 2 shows the timing residuals as a function of frequency, used for a precise determination of the DM value. For this measure we used observations at three different frequencies because the delay in the ToAs inversely depends on the square of the observing frequency as

$$\Delta t = \frac{e^2}{2\pi m_e c}\mathrm{DM}\left(\frac{1}{f_{min}^2} - \frac{1}{f_{max}^2}\right). \tag{4}$$

From Eq. (4), knowing from the observations the time delay in the ToAs (Δt) and the observing frequencies (f_{min} and f_{max}), we can measure the DM value. The larger the $f_{min} - f_{max}$ difference, the more precise the DM determination is.

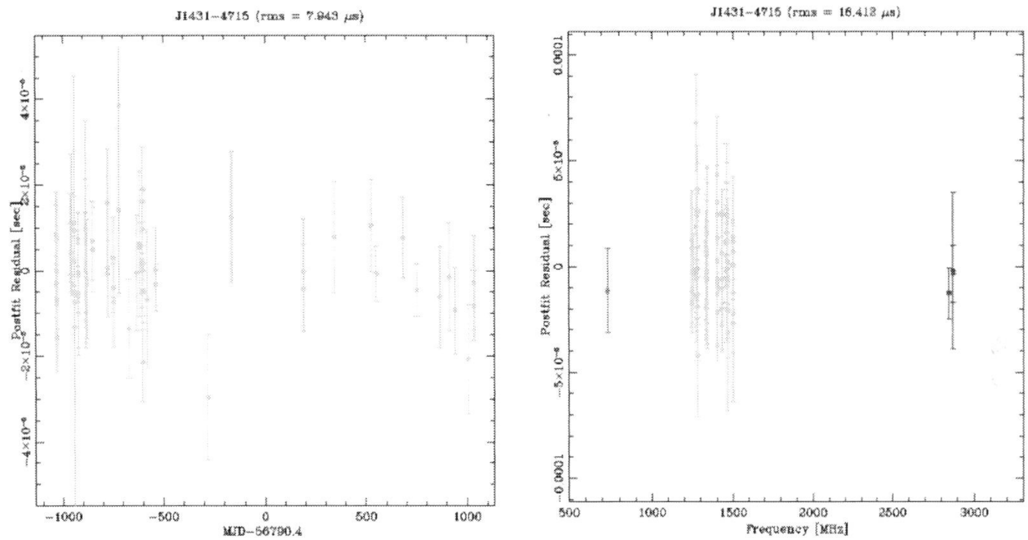

Figure 2. Left plot: timing residuals of J1431−4715 for observations at 1.4 GHz when fitting for all the parameters but the DM. Right plot: timing residuals of J1431−4715 with observations at 0.7 GHz (red), 1.4 GHz (green) and 3.1 GHz (blue) when fitting for the DM.

3. Results

From the measured parameters we can also determine, with some assumptions, other characteristics of the system. One of these is the mass function, defined as

$$f(M_p) = \frac{4\pi^2}{G}\frac{x_p^3}{P_b^2} = \frac{(M_c \sin i)^3}{(M_p + M_c)^2}, \tag{5}$$

where G is the gravitational constant, x_p is the projected semi-major axis of the pulsar orbit, P_b is the orbital period, M_p is the mass of the pulsar, M_c is the companion's mass and i is the inclination of the system (i.e. the angle between the line of sight and the normal to the orbital plane). From the mass function we estimated the minimum companion mass to be 0.13 $M_\odot$ (see Tab. 1), assuming $M_p = 1.4\ M_\odot$ and $i = 90°$.

Finally, assuming that the emission of the pulsar is solely due to the rotational energy loss in the form of magneto-dipole radiation, we estimated the surface dipolar magnetic field, B_0, and the so-called characteristic age, τ_c, as

$$B_0 \approx 3.2 \times 10^{19}\mathrm{G}\sqrt{P\dot{P}}, \qquad \tau_c \approx \frac{P}{2\dot{P}}. \tag{6}$$

The values found are: $B_0 = 1.69 \times 10^8$ G, $\tau_c = 2.25$ Gyr. Some parameters (namely μ_α, μ_δ, DM, x_p and $\dot{P}$) are now measured with higher precision compared to previously published values (see Tab. 1). The other parameters have roughly the same errors of [11].

Pulsar	J1431−4715
Right Ascension, α (J2000)	14:31:44.6182(1)
Declination, δ (J2000)	−47:15:27.567(2)
Proper Motion in α, μ_α (mas yr^{-1})	−11.0(8)
Proper Motion in δ, μ_δ (mas yr^{-1})	−14.2(8)
Spin Frequency, f (s^{-1})	497.02939385196(6)
1st Spin Frequency derivative, $\dot{f}$ (Hz s^{-2})	$-3.4949(7)\times10^{-15}$
Reference Epoch (MJD)	55756.100
Start of Timing Data (MJD)	55758.264
End of Timing Data (MJD)	57845.655
Dispersion Measure, DM (pc cm^{-3})	59.345(2)
Terrestrial Time Standard	TT(TAI)
Time Units	TCB
Number of ToAs	90
Binary Parameters	
Binary Model	ELL1
Projected Semi-major Axis, x_p (lt-s)	0.550053(6)
Epoch of passage at Ascending Node, T_{asc} (MJD)	55756.1047756(4)
Orbital Period, P_b (days)	0.4497391416(7)
Orbital period derivative, $\dot{P}_b$ (10^{-11} s s^{-1})	−1.09(7)
Derived Parameters	
Spin Period, P (s)	$2.01195344253 19(2)\times10^{-3}$
1st Spin Period derivative, $\dot{P}$ (s s^{-1})	$1.4147(3)\times10^{-20}$
Mass Function, $f(M_p)$ ($M_\odot$)	8.83×10^{-4}
Minimum companion mass, $M_{c,min}$ ($M_\odot$)	0.13
Surface Magnetic Field, B_0, (10^8 G)	1.69
Characteristic Age, τ_c (Gyr)	2.25

Table 1. Timing parameters for pulsar J1431−4715. All the uncertainties are the nominal 1-σ values calculated by TEMPO2. ELL1 is a binary orbital model particularly suitable for very-low eccentricities [10].

4. Study of the eclipse region

The results in the previous section do not take into account the observations taken during the eclipse phase. The left plot of Fig. 3 shows the timing residuals for all the observations at three different frequencies.

In the orbital phase range between 0.1 and 0.4 (i.e. around superior conjunction, which occurs at phase 0.25) there is a clear delay in the radio signal. Nevertheless, the radio signal never disappears across the eclipse region. The eclipse lasts for about 2.7 h, corresponding to 25% of the orbital period.

We can notice that the delays depend upon observing frequency and in some observations at 3.1 GHz (blue dots) they show smaller delays than 1.4 GHz (green dots).

We assume that the responsible for the eclipse is a low-density highly-ionized gas cloud enwrapping the companion star.

In this case we can suppose that the delay is due to an orbital phase dependent excess of material between the pulsar and the Earth. Indeed, if we convert the time delays into electron column densities (see Fig. 3, right plot), we can see that the difference between observations at 3.1 GHz and 1.4 GHz is removed. We can conclude that these delays are mainly due to additional DM. The associated extra DM does not cause a decrease in the total flux density, but in addition to the delay in the ToAs, it causes smearing. This phenomenon determines a broadening of the pulse profile due to the rapid variation of the dispersion measure in the eclipse region. The extra contribution $\Delta\mathrm{DM}(\phi)$ to the dispersion measure can be expressed as

$$\Delta\mathrm{DM}(\phi) = \int_{cloud} n_e dl, \tag{7}$$

where the integral is along the gas cloud at the particular orbital phase.

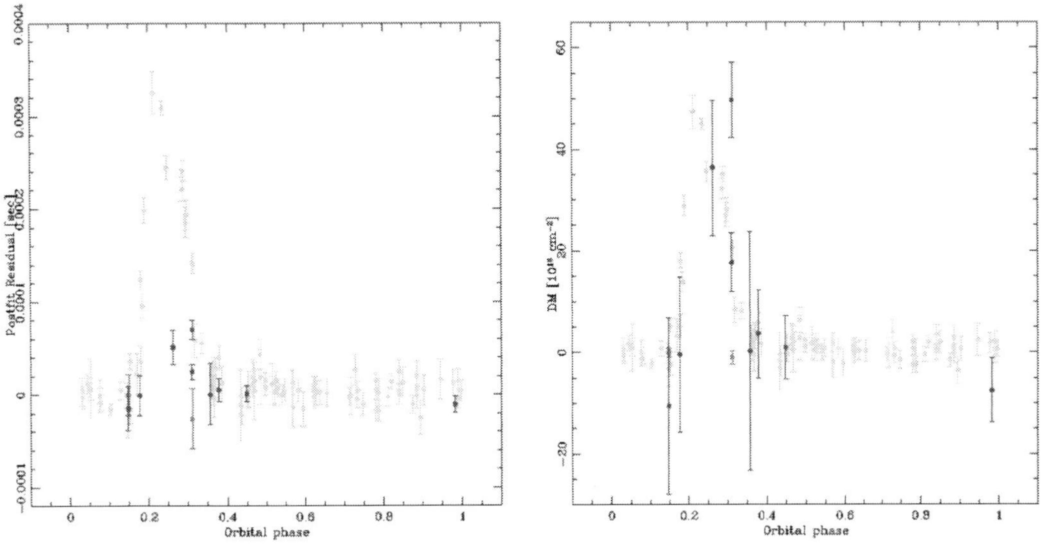

Figure 3. Left plot: ToAs residuals versus orbital phase for J1431−4715 as acquired over a time span of 5 years. The observations were made at 0.7 GHz (red), 1.4 GHz (green) and 3.1 (blue) GHz. Right plot shows the same delays converted to an electron column density.

We can conclude that, as we do not observe a variation in the total flux density, the eclipse in PSR J1431−4715 is not due to absorption processes. Then, the delay is due to the variation of the electron density in the cloud that surrounding the companion star, since that if the delays are converted to an electron column density, the differeces between the delays at 1.4 GHz and 3.1 GHz are removed.

References

[1] Alpar M A, Cheng A F, Ruderman M A and Shaham J 1982 *Nature* **300** 728
[2] Papitto A, Ferrigno C, Bozzo E *et al* 2013 *Nature* **501** 517
[3] van den Heuvel E P J and van Paradijs J 1988 *Nature* **334** 227
[4] Burderi L, D'Antona F, Menna M T *et al* 2001 *X-ray Astronomy 2000* **234** 237
[5] Keith M J, Jameson A, van Straten W *et al* 2010 *MNRAS* **409** 619
[6] Sarkissian J M, Carretti E and van Straten W 2011 *American Institute of Physics Conference Series* **1357** 351
[7] Lorimer D R, Kramer M 2005 *Cambridge University Press* **"Handbook of pulsar Astronomy"**
[8] Hotan A W, van Straten W and Manchester R N 2004 *Publ. Astron. Soc. Australia* **21** 302
[9] Hobbs G B, Edwards R T and Manchester R N 2006 *MNRAS* **369** 655
[10] Lange C, Camilo F, Wex N *et al* 2001 *MNRAS* **326** 274
[11] Bates S D, Thornton D, Bailes M *et al* 2015 *MNRAS* **446** 4019

IOP Conf. Series: Journal of Physics: Conf. Series **956** (2017) 012005 doi:10.1088/1742-6596/956/1/012005

A general approach for testing non-cold dark matter at small cosmological scales

R Murgia

SISSA and INFN, Sezione di Trieste, via Bonomea 265, 34136 Trieste, Italy

E-mail: riccardo.murgia@sissa.it

Abstract. We present a general approach for modelling the small-scale suppression in the linear matter power spectrum induced by the presence of non-cold dark matter. We show that the new parametrisation accurately describes a large variety of non-thermal scenarios, removing the need to individually test each of them. We discuss the first astrophysical constraints on its free parameters and we outline the next steps for pursuing a full statistical data analysis.

1. Introduction

According to the standard cosmological model (ΛCDM model), the universe today is mainly composed by a cosmological constant (Λ) and by cold dark matter (CDM). While being in excellent agreement with cosmic microwave background and large-scale structure observations, this paradigm shows some tensions with structure formation data at sub-galactic scales, often denoted as the CDM "small-scale crisis". Assuming the standard cosmological model, indeed, N-body simulations predict too many dwarf galaxies within the Milky Way (MW) virial radius (*missing satellite* problem [1, 2]) and too much dark matter (DM) in the innermost regions of galaxies (*cusp-core* problem [3]), with respect to observations. Furthermore, the dynamical properties of the most massive MW satellites are not correctly predicted by simulations (*too-big-to-fail* problem [4, 5]). These discrepancies may be relaxed either by baryon physics, still not perfectly included in cosmological simulations [6, 7], or by modifying the standard CDM framework, given that the fundamental nature of DM is still unknown [8].

DM candidates are generally categorised according to their velocity dispersion, which defines a free-streaming length. On scales smaller than their free-streaming length, density fluctuations are erased and gravitational clustering is suppressed. The velocity dispersion of CDM candidates is by definition so small that the corresponding free-streaming length does not have any influence on cosmological structure formation. On the other hand, various non-cold DM (nCDM) scenarios predict structure formation to be suppressed at small cosmological scales and thus have been studied as a viable solution for the small-scale crisis (e.g., sterile neutrinos [9, 10], ultralight scalars [11–13], mixed (cold + warm) DM fluids [8, 14], Self-Interacting DM (SIDM) [15, 16]).

The suppression in the matter power spectrum induced by nCDM can be characterised by different strength and shape, depending on the fundamental nature of the DM candidate. Up to now, lots of efforts have gone into exploring the astrophysical consequences of thermal Warm DM (WDM) models, i.e. candidates with a Fermi-Dirac or Bose-Einstein momentum distribution, which implies a very specific shape of the small-scale suppression, only depending on the WDM particle mass [17, 18]. However, most of the nCDM candidates listed above, well motivated by theoretical particle physics, do not feature a thermal momentum distribution: this can yield to non-trivial suppressions in their power spectra, not adequately captured by the thermal WDM case [19, 20].

In this paper we present a general analytical fitting formula which accurately reproduces the small-scale power suppression induced by the most viable (non-thermal) nCDM scenarios: sterile neutrinos, mixed (cold + warm) fluids, ultralight scalar DM, as well as other models suggested by effective theory of structure formation (ETHOS). We also discuss the first, preliminary constraints on its three free parameters from Lyman-α forest data, based on linear theory. We finally sketch the next steps in order to perform a more comprehensive data analysis.

2. A new general approach

The small-scale suppression of the matter power spectrum $P(k)$, due to the existence of nCDM, is usually described by the transfer function $T(k)$. It is defined as follows:

$$T^2(k) = \left[\frac{P(k)_{\rm nCDM}}{P(k)_{\rm CDM}}\right], \tag{1}$$

i.e. the square root of the ratio of the linear power spectrum in the presence of nCDM with respect to that in the presence of CDM only, for fixed cosmological parameters. For the particular case of thermal WDM, the transfer function may be approximated by the analytical fitting function [21]

$$T(k) = [1 + (\alpha k)^{2\mu}]^{-5/\mu}, \tag{2}$$

where α is the only free parameter and $\mu = 1.12$. Therefore, bounds on the mass of the thermal WDM candidate are easily converted into constraints on α, through the following formula [14]:

$$\begin{aligned} \alpha &= 0.24 \left(\frac{m_x/T_x}{1\ \mathrm{keV}/\mathrm{T}_\nu}\right)^{-0.83} \left(\frac{\omega_x}{0.25(0.7)^2}\right)^{-0.16} \mathrm{Mpc} \\ &= 0.049 \left(\frac{m_x}{1\ \mathrm{keV}}\right)^{-1.11} \left(\frac{\Omega_x}{0.25}\right)^{0.11} \left(\frac{h}{0.7}\right)^{1.22} h^{-1}\mathrm{Mpc}\,, \end{aligned} \tag{3}$$

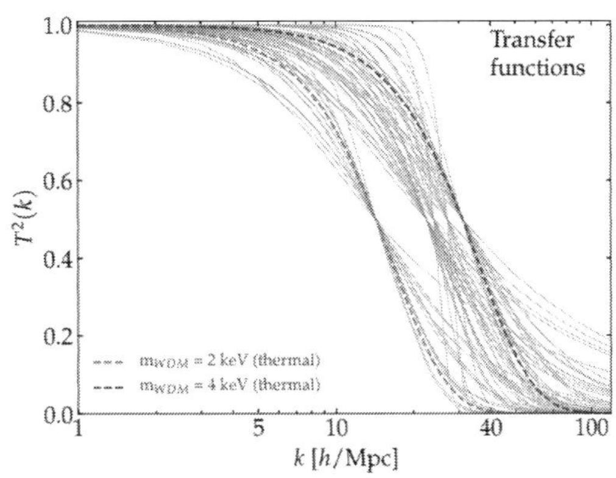

Figure 1. Here we report the plot of the 55 transfer functions, computed through Eq. (4) and associated with the $\{\alpha, \beta, \gamma\}$-combinations considered in the analyses performed by the authors of [19]. The red and blue dashed lines represent the "old" transfer functions (computed via Eq. (2)) for $m'_x = 2$ keV and $m'''_x = 4$ keV, respectively.

with m_i being the mass, T_i the temperature, Ω_i the abundance of the i-th species and $\omega_i \equiv \Omega_i h^2$. The index $i = x, \nu$ stands for WDM and active neutrinos, respectively.

Let us now define the half-mode scale, $k_{1/2}$, as the wave-number for which $T^2 \equiv 0.5$, and introduce the following generalisation of Eq. (2) [19]:

$$T(k) = [1 + (\alpha k)^{\beta}]^{\gamma}, \tag{4}$$

so that $k_{1/2}$ is a function of the three parameters α, β and γ, i.e.

$$k_{1/2} = ((0.5)^{1/2\gamma} - 1)^{1/\beta})\alpha^{-1}. \tag{5}$$

Via Eqs. (2) and (3) we have a one-to-one correspondence between m_x and α. On the other hand, through Eqs. (4) and (5), bounds on the DM mass are mapped to 3D surfaces in the $\{\alpha, \beta, \gamma\}$-space. In other words, given a value of $k_{1/2}$ which corresponds to a certain thermal WDM mass, Eq. (5) allows to compute the corresponding surface in a 3D parameter space.

Recent analyses claim that thermal warm DM candidates with masses of the order of 3 keV can induce a suppression in the corresponding matter power spectra such that the CDM small-scale crisis vanishes or it is largely reduced [22, 23]. It is thus interesting to explore the volume of the $\{\alpha, \beta, \gamma\}$-space associated to thermal WDM masses between 2 and 4 keV by building a 3D grid in the parameter space which samples that volume. Each of the grid points is univocally identified by a certain $\{\alpha, \beta, \gamma\}$-combination corresponding to a different nCDM model.

In Fig. 1 we plot 55 transfer functions, computed through Eq. (4) and associated with the $\{\alpha, \beta, \gamma\}$-combinations considered in the analyses performed by the authors of [19].

Let us stress that the position of the half-mode scale $k_{1/2}$ is still set by the value of α, even in the new general parametrisation, while β and γ are responsible for the shape of the transfer functions before and after $k_{1/2}$, respectively. β has to be greater than zero in order to have meaningful transfer functions, since $\beta < 0$ gives a $T(k)$ that differs from 1 at large scales. The larger is β, the flatter is the transfer function before $k_{1/2}$. Analogously, the larger is $|\gamma|$, the sharper is the cut-off on small scales.

Fig. 1 clearly shows that the new fitting formula is flexible enough to disentangle even tiny differences in the shape of the power suppression and, thus, to discriminate between distinct (non-thermal) nCDM models with power spectra suppressed at very similar scales.

3. The "area criterion" for the Lyman-α forest

We now present a simple method, based on linear theory, for testing different DM scenarios with the Lyman-α forest [24], which is produced by the absorption of the inhomogeneous distribution of the intergalactic neutral hydrogen along different line of sights to distant quasars [25]. Thereby, it constitutes a key observable for efficiently investigating the matter power spectrum at small cosmological scales, namely 0.5 Mpc/$h \lesssim \lambda \lesssim$ 100 Mpc/h [17, 18].

Absolute constraints on DM properties can be extracted from Lyman-α forest data only through a comprehensive statistical analysis. However, it is possible to look into deviations with respect to the most updated limits on the mass of thermal WDM candidates, obtained with a recent thorough analysis [18]. This can be done by applying the "area criterion" introduced in [19], which we briefly summarise in the next paragraphs.

The deviation of a model with respect to the standard CDM case is parameterised by the ratio

$$r(k) = \frac{P_{\rm 1D}(k)}{P_{\rm 1D}^{\rm CDM}(k)}, \tag{6}$$

where $P_{1D}(k)$ is the 1D power spectrum of the model that we are considering, computed by the following integral on the 3D linear matter power spectrum, $P(k')$, at redshift $z = 0$:

$$P_{1D}(k) = \frac{1}{2\pi} \int_k^\infty dk' k' P(k'). \tag{7}$$

In order to find out whether a model deviates more or less from the standard CDM case, with respect to the reference model that we have chosen, we adopt the following criterion: a model is rejected if it shows a larger power suppression with respect to the reference one. The suppression in the power spectra is computed via the following estimator:

$$\delta A \equiv \frac{A_{\rm CDM} - A}{A_{\rm CDM}}, \tag{8}$$

where A is the integral of $r(k)$ over the range of scales probed by Lyman-α observations ($0.5\ h/\text{Mpc} < k < 20\ h/\text{Mpc}$ for the MIKE/HIRES+XQ-100 dataset used in [18]), i.e.

$$A = \int_{k_{\rm min}}^{k_{\rm max}} dk\ r(k), \tag{9}$$

so that $A_{\rm CDM} \equiv k_{\rm max} - k_{\rm min}$, by definition.

The area criterion that we have sketched above can now be applied for constraining the $\{\alpha, \beta, \gamma\}$ parameter space. We take as reference $m_{\rm WDM} = 5.3$ keV (2σ C.L.), which is the most stringent lower limit on thermal WDM masses from Lyman-α forest data up to date [18]. By plugging the linear power spectrum associated to this reference model into Eqs. (6) and (8), we find $\delta A_{\rm REF} = 0.21$, which is the estimate of the small-scale power suppression for nCDM models that are excluded at 2σ C.L. by Lyman-α forest data. Analogously, we compute the same estimator δA in a grid in the $\{\alpha, \beta, \gamma\}$-space, where each grid point corresponds to a distinct nCDM model, and we accept (at 2σ C.L.) only those $\{\alpha, \beta, \gamma\}$-combinations that display a suppression $\lesssim 21\%$ with respect to the CDM power spectrum (i.e. those models for which $\delta A < \delta A_{\rm REF}$).

In Fig. 2 we show a 3D contour plot, which represents the volume of the $\{\alpha, \beta, \gamma\}$-space that contains models in agreement with Lyman-α forest data, according to the area criterion.

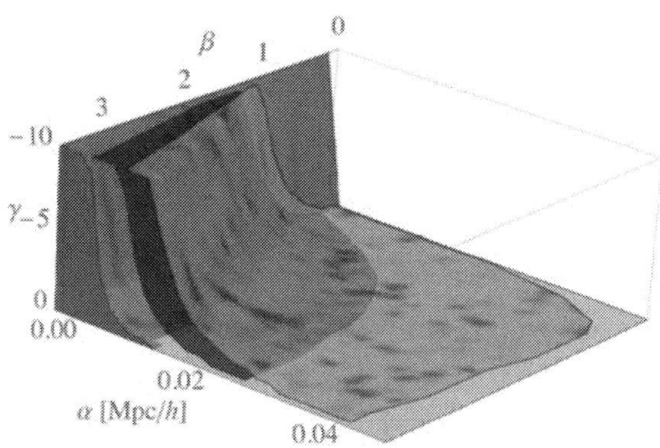

Figure 2. Here we show a 3D contour plot in the $\{\alpha, \beta, \gamma\}$-space (from [19]) which represents the volume of the parameter space in agreement with Lyman-α forest data, according to the area criterion. The red contour is the 1σ C.L. limit, while the blue and green contours are the 2σ and 3σ C.L. constraints, respectively. All those models associated to $\{\alpha, \beta, \gamma\}$-triplets sampling the non-coloured region are thereby excluded at 3σ C.L. by our analysis.

4. Connection with theoretical particle physics models

We now aim to highlight the link between the fundamental nature of DM and our 3D parametrisation. To do so, one should compare the predictions in terms of structure formation in order to determine to which extent the 3-parameter fitting formula matches the "true" transfer functions associated with various viable nCDM scenarios provided by theoretical particle physics. By doing this, we can figure out whether the $\{\alpha, \beta, \gamma\}$-fit to a certain nCDM model leads us to the same conclusion about its validity when confronted with the Lyman-α forest constraints, computed by applying the area criterion.

The authors of [19] have performed this analysis by considering most of the known nCDM particle scenarios: resonantly produced (RP) sterile neutrinos, sterile neutrinos by particle decays, cold + warm DM fluids, ultralight scalar DM and ETHOS models. For each class of models, 5 different $\{\alpha, \beta, \gamma\}$-combinations corresponding to different properties of the given scenario (e.g. different DM masses, lepton asymmetry, nCDM abundance) have been selected and compared with the analogous actual transfer functions. The main result is that the $\{\alpha, \beta, \gamma\}$-fit predictions depart from the "true" results by few per cent at most, so that the fitted transfer functions practically always lead to the same conclusion provided by the actual models.

Therefore, whenever one wants to test a nCDM particle scenario with structure formation data, it is sufficient to match the resulting transfer functions to Eq. (4) and check whether the fitted 3D points are allowed, i.e. whether they sample the coloured volume of the parameter space shown in Fig. 2. However, being the results reported in [19] based on linear theory only, they must be considered as a first step in the direction of a more comprehensive analysis. This analysis will be based on a large suite of high resolution hydrodynamical simulations, in order to extract more accurate constraints from Lyman-α forest data. In the next section we briefly outline the path towards such extensive study.

5. Towards a full Monte Carlo Markov Chain analysis

The physical observable for Lyman-α forest experiments is the flux power spectrum, $P_{\rm F}(k, z)$, rather than the 1D (or 3D) linear matter power spectrum. Nevertheless, two distinctive characteristics of the Lyman-α physics suggest that the area criterion analysis could be also quantitatively correct. Firstly, the use of Eq. (6) with flux power spectra is justified by the existing relation between matter and flux power spectra, namely $P_{\rm F} = b^2(k)P_{\rm 3D}(k)$, where the bias factor $b^2(k)$ differs very little for models reasonably close to the standard CDM scenario [14]. Moreover the area criterion is motivated by the fact that the peculiar velocities of the intergalactic medium (generally < 100 km/s) tend to spread the small-scale power within a relatively broad range of wave-numbers in the explored region [26].

It is thus worth to test the accuracy of the area criterion against a full statistical data analysis. This has been done for the first time in [27], in the context of ultralight scalar DM, i.e. in a 2D parameter space where different models are identified by different combinations of scalar DM mass and abundance $\{m, F\}$. The authors of [27] have performed both the area criterion investigation and a full Monte Carlo Markov Chain (MCMC) analysis of the $\{m, F\}$-space, finding agreement at 10% level between the 2σ contours resulting from the two different methods.

Therefore, given that the area criterion represents a simple and intuitive, yet approximate, method for constraining the parameter space of nCDM models, a more precise and extensive approach, not limited to linear theory, is needed. Indeed, we are currently developing a MCMC code which samples the $\{\alpha, \beta, \gamma\}$-volume enclosed by a non-regular grid of 55 full hydrodynamical simulations, each of them associated to a distinct combination of the three parameters, in order to compare the corresponding flux power spectra with the actual physical observables provided by Lyman-α forest experiments. This will allow to determine accurate absolute bounds on the three parameters of Eq. (4).

6. Conclusions

In spite of strong efforts both in particle physics and cosmology, the fundamental nature and composition of DM remain undiscovered. Nevertheless, the majority of the constraints on nCDM properties currently available only apply to the very specific shape of the power suppression corresponding to the thermal WDM case.

Given that most of the viable DM candidates provided by theoretical particle physics are not adequately described by the oversimplified concept of thermal WDM, we have presented a new analytical fitting formula for the transfer function $T(k)$ [19]. Due to the mutual dependence among its three free parameters, the new general formula is capable to reproduce a huge variety of shapes in the suppression of the linear matter power spectrum. We have pointed out that the new parametrisation is able to embrace all the most viable nCDM particle scenarios, such as sterile neutrinos, cold + warm DM fluids, ultralight scalars and DM models from effective theory of structure formation (ETHOS). We have discussed the first constraints on the three free parameters characterising the new approach, obtained by applying an approximate yet very intuitive method, based on linear theory, i.e. the area criterion for Lyman-α forest data [19, 27]. The Lyman-α forest constitutes, in fact, a powerful tool for constraining DM properties down to very small cosmological scales.

The results discussed here represent a first step towards a fully general modelling of the small-scale departures from the standard CDM model, which will include an accurate and extensive MCMC analysis of Lyman-α forest data, providing absolute limits easily translatable to bounds on the fundamental nCDM properties through the scheme that we have illustrated.

References

[1] Klypin A *et al* 1999 *Astrophys. J.* **522** 82–92.
[2] Moore B *et al* 1999 *Astrophys. J.* **524** L19–L22
[3] de Blok W 2010 *Advances in Astronomy* **2010** 789293
[4] Boylan-Kolchin M *et al* 2011 *Mon. Not. Roy. Astron. Soc.* **415** L40–L44
[5] Boylan-Kolchin M *et al* 2012 *Mon. Not. Roy. Astron. Soc.* **422** 1203–18
[6] Okamoto T *et al* 2008 *Mon. Not. Roy. Astron. Soc.* **390** 920
[7] Governato F *et al* 2012 *Mon. Not. Roy. Astron. Soc.* **422** 1231–40
[8] Schneider A *et al* 2017 *Mon. Not. Roy. Astron. Soc.* **470** 1542
[9] Adhikari R *et al* 2017 *JCAP* **1701(01)** 025
[10] König J *et al* 2016 *JCAP* **1611(11)** 038
[11] Hu W *et al* 2000 *Phys. Rev. Lett.* **85** 1158–61
[12] Marsh D and Silk J 2014 *Mon. Not. Roy. Astron. Soc.* **437(3)** 2652–63
[13] Hui L *et al* 2017 *Phys. Rev. D* **95(4)** 043541
[14] Viel M *et al* 2005 *Phys. Rev. D* **71(6)** 1–10
[15] Cyr-Racine F *et al* 2016 *Phys. Rev. D* **93(12)** 123527
[16] Vogelsberger M *et al* 2016 *Mon. Not. Roy. Astron. Soc.* **460** 1399
[17] Viel M *et al* 2013 *Phys. Rev. D* **88** 043502
[18] Irsic V *et al* 2017 *Phys. Rev. D* **96(2)** 023522
[19] Murgia R *et al* 2017 *Preprint* astro-ph.CO/1704.07838
[20] Baur J *et al* 2017 *Preprint* astro-ph.CO/1706.03118
[21] Bode P *et al* 2001 *Astrophys. J.* **556(1)** 93–107
[22] Lovell M *et al* 2016 *Mon. Not. Roy. Astron. Soc.* **461(1)** 60–72
[23] Lovell M *et al* 2017 *Mon. Not. Roy. Astron. Soc.* **468(3)** 2836–49
[24] McQuinn M 2016 *Ann. Rev. Astron. Astrophys.* **54** 313
[25] Viel M *et al* 2002 *Mon. Not. Roy. Astron. Soc.* **329** 848
[26] Gnedin N and Hamilton A 2002 *Mon. Not. Roy. Astron. Soc.* **334** 107–116
[27] Kobayashi T *et al* 2017 *Preprint* astro-ph.CO/1708.00015

IOP Conf. Series: Journal of Physics: Conf. Series **956** (2017) 012006 doi:10.1088/1742-6596/956/1/012006

Remote sensing of the solar photosphere: a tale of two methods

G Viavattene[1], F Berrilli[1], M Collados Vera[2], D Del Moro[1], L Giovannelli[1], B Ruiz Cobo[2] and F Zuccarello[3]

[1] Università di Roma "Tor Vergata", Via della Ricerca Scientifica, I-00133 Roma, Italy
[2] Instituto de Astrofisica de Canarias, Calle Vìa Làctea, s/n, 38205 San Cristòbal de La Laguna, Tenerife, Spain
[3] Università degli Studi di Catania, Dipartimento di Fisica e Astronomia, Via Santa Sofia 78, 95123 Catania, Italy

E-mail: giorgio.viavattene@roma2.infn.it

Abstract. Solar spectro-polarimetry is a powerful tool to investigate the physical processes occurring in the solar atmosphere. The different states of polarization and wavelengths have in fact encoded the information about the thermodynamic state of the solar plasma and the interacting magnetic field. In particular, the radiative transfer theory allows us to invert the spectro-polarimetric data to obtain the physical parameters of the different atmospheric layers and, in particular, of the photosphere. In this work, we present a comparison between two methods used to analyze spectro-polarimetric data: the classical Center of Gravity method in the weak field approximation and an inversion code that solves numerically the radiative transfer equation. The Center of Gravity method returns reliable values for the magnetic field and for the line-of-sight velocity in those regions where the weak field approximation is valid (field strength below 400 G), while the inversion code is able to return the stratification of many physical parameters in the layers where the spectral line used for the inversion is formed.

1. Introduction

The interaction between magnetic fields and convection in the solar photosphere regulates a large variety of physical processes that we can observe from Earth with ground-based solar telescopes. One of the challenges of the modern solar physics is to measure very precisely the magnetic field and the thermodynamical properties of the solar photosphere exploiting the polarization of the light, which encodes all this information. In fact, by analyzing the polarization state of the electromagnetic radiation at different wavelengths, we are able to extract the magnetic fields in some layers of the solar photosphere harnessing the radiative transfer and the Zeeman effect theories. Therefore, the Radiative Transfer Equation (RTE) is at the core of this approach. There are different methods and approximations to solve it: there are analytical solutions of the RTE based on several assumptions and there are numerical solutions which, using more realistic atmospheric models, return more information about the physical parameters of the atmosphere. In this work, we present some results obtained using two different methods, applied to interpret spectro-polarimetric observations of the so-called Quiet Sun [1]. We compare the results from the Center of Gravity (CoG) method [2–4], based on weak field approximation and an analytical solution of the RTE, and the NICOLE inversion technique [5,6] which solves numerically the RTE.

IOP Conf. Series: Journal of Physics: Conf. Series **956** (2017) 012006 doi:10.1088/1742-6596/956/1/012006

2. The radiative transfer

The propagation of the electromagnetic radiation through a stellar atmosphere is a complex, non-linear, three dimensional and time dependent problem, that involves the properties of the whole atmosphere. The RTE describes how the radiation propagates through the solar atmosphere:

$$\frac{d\mathbf{I}}{d\tau_c} = \mathbf{K}(\mathbf{I} - \mathbf{S}) \tag{1}$$

where $\mathbf{I} = (I, Q, U, V)^T$ is the Stokes polarization vector (I is the intensity, Q and U are connected to the linear polarization, and V is connected to the circular polarization), $\tau_c = \int_s^{s_{lim}} \chi_{cont}\, ds$ is the continuum optical depth (χ_{cont} is the continuum opacity and s_{lim} is the geometrical distance corresponding to $\tau_c = 0$), $\mathbf{K}$ is the 4×4 propagation matrix and $\mathbf{S}$ is the source function. In the Non Local Thermodynamic Equilibrium (NLTE) approach the RTE is a set of four coupled differential equations and, in order to solve the problem, we have to consider both the radiative and the collisional transitions. Only numerical solutions exist and they require an iterative procedure to be found. In the following, we examine two approximations that are used to solve analytically the RTE.

2.1. The Milne-Eddington approximation

In the Milne-Eddington (ME) approximation the propagation matrix is constant with respect to the optical depth and it is the same at all the wavelengths, while the source function depends linearly on the continuum optical depth. A constant propagation matrix implies that all the parameters that define the spectral line formation (i.e. the line-to-continuum absorption coefficient, the Doppler width and the damping parameter) are constant and also that the magnetic field vector $\mathbf{B}$ and the velocity along the line-of-sight v_{los}, are constant through the atmosphere. Therefore, in the ME approximation the physical parameters are not stratified along the atmosphere. In this approximation the $\mathbf{K}$ matrix, and so the analytical solution, are connected with just nine parameters of the atmosphere: three for the magnetic field vector (the intensity, the inclination along the line-of-sight and the azimuth angle); two for the source function (the constant and the linear terms); three for the shape of the spectral line (the line-to-continuum absorption coefficient, the Doppler width and the damping parameter); and one for the velocity along the line-of-sight v_{los}. Thus, assuming these nine parameters and using the RTE, we can obtain *synthetic* Stokes profiles (i.e. the full polarization state with respect to wavelength). On the contrary, starting from the observed Stokes profiles, we can extract the atmospheric parameters, by the process called *inversion*. It is worth stressing that such a simple atmosphere produces only very simple Stokes profiles.

2.2. The weak field approximation and the Center of Gravity method

Another approximation to solve the RTE is the weak field approximation, which assumes the magnetic field to be constant with respect to the optical depth and below a few hundreds of Gauss. Under such hypotheses, the amplitude of the Stokes V lobes is proportional to the magnetic field intensity. The CoG method [2–4] assumes these hypotheses and allows us to evaluate the component of the magnetic field intensity along the line-of-sight (longitudinal component) using this equation:

$$B_{||} = B\cos\chi = \frac{1.071 \times 10^9}{\bar{g}\lambda_0^2}(\lambda_+ - \lambda_-) \tag{2}$$

where $\bar{g}$ is the effective Landé factor, λ_0 is the laboratory wavelength of the observed spectral line, and λ_+ and λ_- are respectively the positions of the positive and negative lobes of a Stokes

V profiles.

3. Inversion techniques

The previous approximations can only describe simple Stokes profiles. Nature is very often more complex than this. In real solar spectro-polarimetric data, asymmetric Stokes profiles are commonly observed and it is necessary to increase the complexity of the atmospheric model in order to match the observations. In fact, the physical parameters of the solar atmosphere vary with the optical depth, and therefore with the geometrical depth. To describe this stratification of the atmospheric parameters, we can discretize the altitude layers of the atmosphere in a grid of point, called *nodes*.
The physical parameters of the solar atmosphere are extracted from the observed Stokes profiles using an inversion technique. In practice, the distance between the observed Stokes profiles and the synthetic ones is computed using a merit function:

$$\chi^2(\mathbf{x}) = \frac{1}{N_f} \sum_{s=0}^{3} \sum_{i=1}^{q} [I_s^{obs}(\lambda_i) - I_s^{syn}(\lambda_i;\mathbf{x})]^2 \, w_{s,i}^2 \tag{3}$$

where $\mathbf{x}$ is a vector containing the whole set of the atmospheric parameters, N_f is the number of nodes used to discretize the model atmosphere (and defines the number of degrees of freedom of the model), s is the index for the four Stokes profiles, i is the index for the wavelengths, and $w_{s,i}$ are weights used to take into account the different signal-to-noise ratios of the Stokes profiles.
Therefore, the purpose of the inversion techniques is to find the synthetic Stokes profiles that best match the observed ones by iteratively minimizing the merit function. These techniques use a stratified atmosphere model and retrieve a best estimate of the solar atmosphere given the measured Stokes parameters and the chosen model. For a more detailed explanation of the operation of the inversion techniques in general, see [7] and [8].

4. Data analysis

The dataset used for the analysis has been acquired on 2006 with the Interferometric BIdimensional Spectropolarimeter (IBIS [9, 10]) located in the Dunn Solar Telescope (DST) at Sacramento Peak, National Solar Observatory (NSO), New Mexico. The spectro-polarimetric dataset of the Quiet Sun consists in observations of the spectral region containing the FeI 6301.5 Å and 6302.5 Å lines, and, pixel by pixel of the Field of View (FoV), the four Stokes profiles are measured by the instrument. The FoV framed by the instrument is 40×40 arcsec^2, which corresponds approximately on the solar surface to 30×30 Mm^2. For more information on the dataset, see [11–13]. In Fig. 1 the spectral scanning of Stokes I and V profiles performed by IBIS instrument are shown.
The dataset has been calibrated with the IBIS pipeline [14]. In Fig. 2 examples of the Stokes I and V maps of the FoV are shown. The spectral position of the Stokes I map is centered in the continuum, while the spectral position of the Stokes V map is centered in the negative lobe of the 6301.5 Å line. In the Stokes I map is clearly visible the photospheric granulation which is associated to the temperature of the solar photosphere; in fact, the central part of the granules is hotter than the surrounding and the intergranular lanes are cooler. In the Stokes V map some coherent magnetic structures are visible. It is interesting to note that these magnetic structures do not exhibit an evident counterpart in the Stokes I map.

The dataset has been analyzed using both NICOLE and CoG methods. The characterizing parameters used for the inversion with NICOLE are: temperature represented with three nodes;

IOP Conf. Series: Journal of Physics: Conf. Series **956** (2017) 012006 doi:10.1088/1742-6596/956/1/012006

v_{los} represented with two nodes; magnetic field constant with height; no micro-turbulence contribution considered; no macro-turbulence considered, but instrumental spectral profile taken into account; initializing atmosphere model Harvard-Smithsonian Reference Atmosphere [15]. The choice of the number of nodes is a trade-off between the quality of the data and the complexity (number of degrees of freedom) of the model atmosphere which can be retrieved by the inversions. Given the spectral sampling, the signal-to-noise ratio and the number of spectral lines of the dataset, our choice grants an adequate atmosphere stratification without data over-fitting.

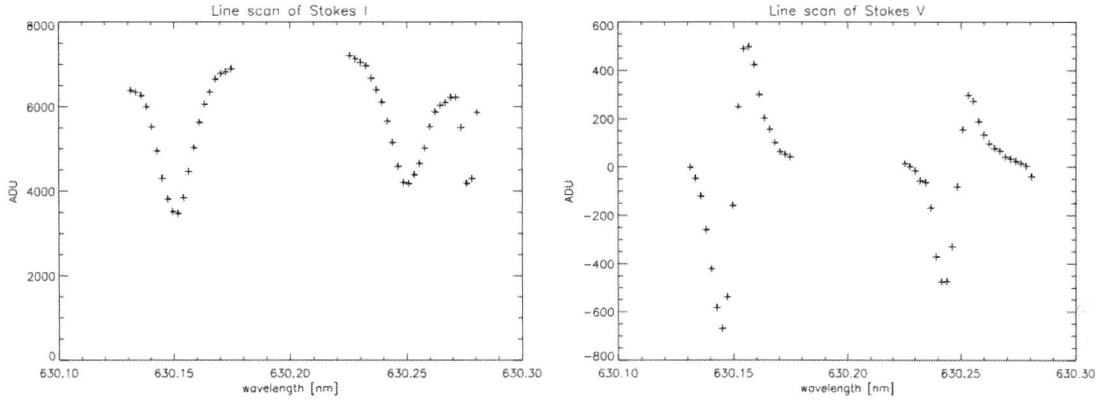

Figure 1. Left panel: IBIS scan of Stokes I profile of Fe I 630.15 nm and 630.25 nm spectral lines. At 630.28 nm is visible a telluric line due to absorption of water in the Earth atmosphere. Right panel: IBIS scan of Stokes V profile of Fe I 630.15 nm and 630.25 nm spectral lines. As expected, the telluric line shows no circular polarization signal. All counts are reported in Analog-Digital Units (ADU).

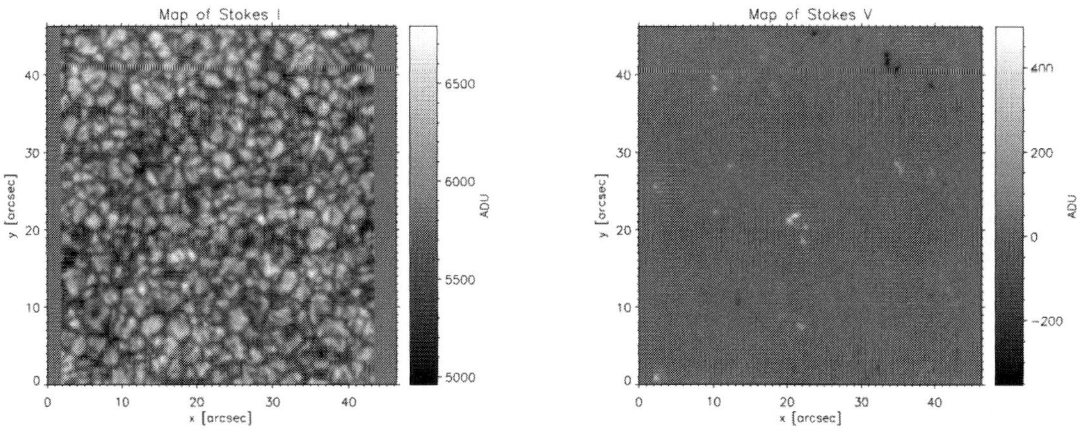

Figure 2. Left panel: Map of Stokes I in the continuum. Right panel: Map of Stokes V in the negative lobe of the 630.15 nm line. Counts are reported in Analog-Digital Units (ADU).

4.1. Comparisons and discussion: the CoG method vs. the inversions

In this subsection we will present the results of the data analysis and some comparisons between the two methods (CoG method and the NICOLE inversion technique). In particular, we focus our attention on the longitudinal component of the magnetic field. For this analysis, the frames with poor seeing are not considered.

An Interactive Data Language (IDL) code has been developed in order to perform the evaluation of the longitudinal component of the magnetic field analyzing the spectro-polarimetric dataset pixel by pixel and using Eq. 2. The values of the longitudinal magnetic field have been obtained also with the NICOLE inversion code. The resulting magnetic field maps of one temporal frame are reported in Fig. 3. It can be noted that both methods retrieve coherent magnetic structures although the analysis has been performed pixel by pixel.

We also compare the Probability Density Functions (PDFs) of the longitudinal magnetic field

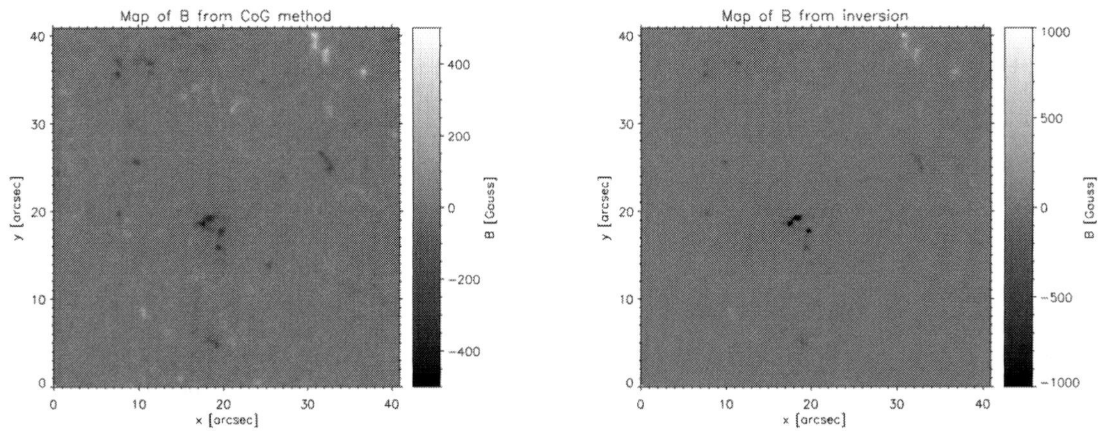

Figure 3. Left panel: Map of the longitudinal component of the magnetic field evaluated with the CoG method. The map is saturated between -500 G (black) and +500 G (white) according to the weak field approximation regime. Right panel: Map of the longitudinal component of the magnetic field evaluated with NICOLE inversions. The map is saturated between -1000 G (black) and +1000 G (white).

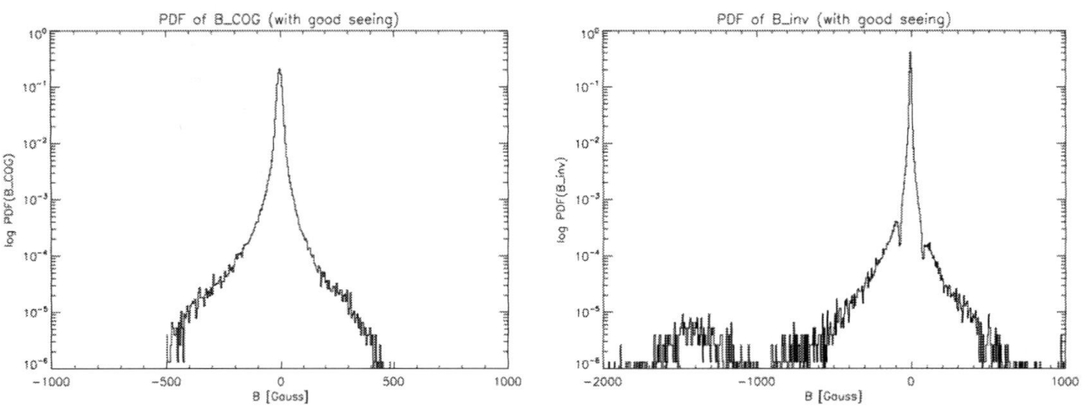

Figure 4. Left panel: PDF of the longitudinal magnetic field evaluated using the CoG method. Right panel: PDF of the longitudinal magnetic field evaluated with inversion techniques.

values retrieved by the two methods. In Fig. 4 (left panel) the PDF of the longitudinal magnetic field evaluated with the CoG method is shown. It can be noted that the wings of the PDF spread out to approximately -500 G and +500 G at most. In fact, for magnetic fields greater than 400 G the weak field approximation fails and the expression in Eq. 2 is not valid anymore, thus producing such a saturation effect in the PDF. Nevertheless, the PDF from the CoG method is quite symmetric. In Fig. 4 (right panel) the PDF of the longitudinal magnetic field evaluated with NICOLE inversions is shown. This PDF is peculiar because, apart from some pixels that have magnetic field values below -1000 G, its central part is not symmetric and two asymmetric bumps are visible. A possible explanation to these features is related to the tendency of inversion techniques to overestimate weak fields: in fact, pixels which should have lower absolute values of magnetic field are shifted by the inversions to higher absolute values, and therefore they contribute to the formation of the two bumps.
In this analysis, pixels with magnetic field values below three times the noise values have been excluded.
Even considering the saturation effect of the CoG method for high magnetic field values, it is evident that the magnetic field retrieved by NICOLE is systematically larger than the one retrieved by the CoG. This effect has also been discussed in [16]. As of today, improving reliability of the inversion retrievals in the weak field range is a very dynamic field of research.

5. Conclusion

We conclude by summarizing the principal advantages and disadvantages of the two methods which have been used to analyze the spectro-polarimetric data.
On one hand, the inversion technique requires long computation time, is able to retrieve the stratification of the physical parameters along the solar atmosphere, but tends to overestimate weak magnetic fields. On the other hand, the CoG method has more assumptions and returns an analytical result in a very short computational time. It retrieves a value of the longitudinal magnetic field (as well as of the line-of-sight velocity) "averaged" along the atmospheric layers which mostly contribute to the observed spectral line shape.

References

[1] Del Moro D 2004 *Astronomy & Astrophysics* **420** 1007
[2] Semel M D 1967 *Annales d'Astrophysique* **30** 513
[3] Rees D E and Semel M D 1979 *Astronomy & Astrophysics* **74** 1R
[4] Landi Degl'Innocenti E and Landolfi M 2004 *Astrophysics and Space Science Library* **307** L
[5] Socas-Navarro H, de la Cruz Rodríguez J, Asenzio Ramos A, Trujillo Bueno J and Ruiz Cobo B 2015 *Astronomy & Astrophysics* **577** 7
[6] Socas-Navarro H 2015 *Astrophysics Source Code Library* ascl:1508.002
[7] Ruiz Cobo B and del Toro Iniesta J C 1992 *The Astrophysical Journal* **398** 375
[8] Ruiz Cobo B and del Toro Iniesta J C 1994 *Astronomy & Astrophysics* **283** 129
[9] Cavallini F, Berrilli F, Cantarano S and Egidi A 2001 *Mem. SAIt* **72** 554
[10] Cavallini F 2006 *Solar Physics* **236** 415
[11] Viticchié B, Del Moro D, Berrilli F, Bellot Rubio L and Tritschler A 2009 *Astrophysical Journal* **700** L145
[12] Del Moro D, Giannattasio F, Berrilli F, Consolini G, Lepreti F and Gošić M 2015 *Astronomy & Astrophysics* **576** A47
[13] Del Moro D, Giovannelli L, Pietropaolo E and Berrilli F 2017 *Experimental Astronomy* **43** 23
[14] Viticchié B, Del Moro D, Criscuoli S and Berrilli F 2010 *The Astrophysical Journal* **273** 787
[15] Gingerich O, Noyes R W, Kalkofen W and Cuny Y 1971 *Solar Physics* **18** 347
[16] del Toro Iniesta J C and Ruiz Cobo B 2016 *Living Review in Solar Physics* **13** 1 4 84 pp.

The MONET code for the evaluation of the dose in hadrontherapy

A Embriaco[1,2]

[1] University of Pavia, Physics Departiment, Via Bassi 6, 27100 Pavia, Italy
[2] INFN Pavia, Via Bassi 6, 27100 Pavia, Italy

E-mail: alessia.embriaco@pv.infn.it

Abstract. The MONET is a code for the computation of the 3D dose distribution for protons in water. For the lateral profile, MONET is based on the Molière theory of multiple Coulomb scattering. To take into account also the nuclear interactions, we add to this theory a Cauchy-Lorentz function, where the two parameters are obtained by a fit to a FLUKA simulation. We have implemented the Papoulis algorithm for the passage from the projected to a 2D lateral distribution. For the longitudinal profile, we have implemented a new calculation of the energy loss that is in good agreement with simulations. The inclusion of the straggling is based on the convolution of energy loss with a Gaussian function. In order to complete the longitudinal profile, also the nuclear contributions are included using a linear parametrization.
The total dose profile is calculated in a 3D mesh by evaluating at each depth the 2D lateral distributions and by scaling them at the value of the energy deposition. We have compared MONET with FLUKA in two cases: a single Gaussian beam and a lateral scan. In both cases, we have obtained a good agreement for different energies of protons in water.

1. Introduction

Hadrontherapy is an oncological technique that uses heavy particles, in particular protons and Carbon ions, for the tumors treatment [1]. The advantage of using heavy particles instead of photons is due to the depth dose profile. In conventional radiotherapy, the photons lose most of their energy in the entrance region. On the contrary, the depth dose profile of heavy particles is relatively small at the beginning and reaches a maximum at the end of their range, followed by a decrease to almost zero values. This feature is known as Bragg Peak [2] and it allows, by setting an appropriate energy, to concentrate the dose to the tumor tissue.
Therefore, the accurate evaluation of the dose distribution is important to delimit the prescribed dose to the cancer volume, saving as much as possible the surrounding healthy tissue.
In this study, we have evaluated the energy deposited of proton beams in water with the new MONET code comparing the results with the Monte Carlo (MC) simulations.

2. The MONET code

MONET (Model of ioN dosE for Therapy) is a fast and accurate code for the evaluation of the 3D dose distribution for protons in water introduced recently by our group [3].
The input variables are the energy of proton and the beam shape: for example in this study we use Gaussian beams of energy 100, 150 and 200 MeV. The first step of the MONET code is the evaluation of lateral distribution taking into account the Multiple Coulomb scattering and

the nuclear interaction. Afterwards, starting from the 1D lateral profile the model reconstructs the 2D radial distribution. The last step is the calculation of energy deposition and the scale of the 2D lateral distribution at this value. In this way, our code is able to evaluate a 3D dose distribution. The free parameters for the nuclear interaction are only four compared with other model [4]: two parameters for the lateral profile (Cauchy-Lorentz function) and two for the linear parametrization in the longitudinal profile.
All results are compared with the FLUKA code, which is currently used in many hadrontherapy centers [5] and has been recently validated with experimental data [6].

2.1. Lateral profile

The first step of the implementation of MONET code is the evaluation of lateral distribution. The shape of the lateral profile comes from the combination of two processes: the multiple Coulomb scattering and nuclear interactions. The MONET code is based on the Molière theory of multiple Coulomb scattering [6, 7] . To take into account also of the nuclear interaction, we have added to Molière distribution the Cauchy-Lorentz function, that describes the nuclear tail of distribution. The total lateral distribution is given by [6]:

$$f_x(x) = W_p f_M(x) + (1 - W_p)\frac{t(x)}{\int t(u)\mathrm{d}u}. \qquad (1)$$

The Cauchy-Lorentz distribution $t(x)$ has only two free parameters, that are obtained by a fit to FLUKA simulation. The Molière distribution is weighted for the primary particle weight W_p, evaluated using the formula present in the Ulmer paper [8], while the Cauchy-Lorentz distribution is multiplied by $1 - W_p$ the nuclear contribution [3].
In Fig. 1 (left), the total lateral distribution is in agreement with FLUKA simulation.

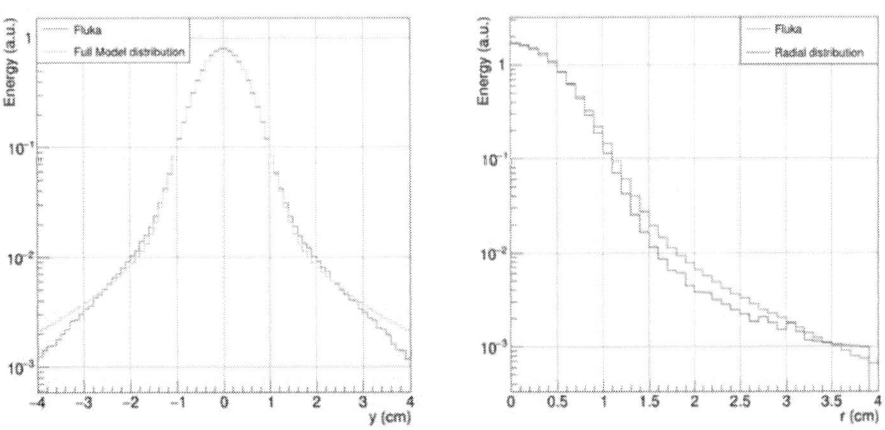

Figure 1. Lateral distributions (left) and radial distributions (right) for 150 MeV protons in water at a depth z=11 cm.

Now, the next step is the passage from the projected lateral distribution to a 2D radial distribution.
The projected distribution f_x and f_y are uncorrelated but not independent [9]. Therefore, in this case is not possible to use the factorization theorem for independent probability distribution. A solution is given by the Papoulis theorem [10] that allows, in case of cylindrical symmetry, to

rebuild the radial distribution starting from the projected one.
We have implemented the Papoulis algorithm in the code and the comparison with simulation is good (Fig.1 right).

2.2. Longitudinal profile

The second part of MONET code is the evaluation of energy loss process in the longitudinal way.
The longitudinal profile is mainly determined by the stopping process of protons due to inelastic collisions with the atomic electrons. In our model, we use the following expression for the average energy loss:

$$E_K(z) = -m + \frac{F(z)}{2} + \sqrt{m^2 + \frac{F^2(z)}{4}}, \qquad F(z) = p\beta\left(1 - \frac{z}{R}\right)^{k/2}, \qquad k = 1.07. \tag{2}$$

Our formula has a similar behaviour to the relation already present in the literature [11, 12]. We have analyzed different therapeutic energies and our formula is in good agreement with FLUKA simulations within 1%.
Due to the statistical nature of the energy loss process, the longitudinal profile is characterized by straggling, that produces a widening of the Bragg peak. A good approximation for the straggling is given by the convolution of the energy loss with a Gaussian function [11, 13]:

$$\hat{E}_K(z)0 = \int_0^R E_K(\bar{z})\frac{e^{-(z-\bar{z})^2/2\sigma^2}}{\sqrt{2\pi}\sigma}d\bar{z}, \qquad \sigma \approx 0.012R^{0.935} \tag{3}$$

The inclusion of straggling effects in our model is still in agreement with simulations.
Also nuclear interactions play an important role by reducing the fluence of the primary protons, in the order about 1% per centimeter in water. In order to complete the longitudinal dose profile, the nuclear contributions are included in the calculation using a linear parametrization where the coefficients are obtained by fitting a FLUKA simulations in the presence of the nuclear interactions.

$$f_z(z) = W_p\hat{E}_K(z) + (1 - W_p)E_N(z) \tag{4}$$

where $E_N(z)$ is a linear parametrization for the nuclear contribution $E_N(z) = az + b$, with the coefficients a and b are obtained by fitting FLUKA simulations.
The average energy loss is weighted for the primary particle weight [8], while the nuclear part is multiplied for nuclear contribution, using the same approach of the lateral profile. In Fig. 2, the comparison between the evaluation of total energy deposition and the FLUKA simulation shows a good agreement (1%).
In conclusion, the MONET code is able to evaluate the total dose profile in a three dimensional mesh with only four free parameters, by calculating at each depth the longitudinal deposited energy and by distributing it laterally on the transverse plane. Therefore the dose is the product between the longitudinal energy deposition and the radial probability distribution in the transverse plane.

3. Results

To assess the accuracy of our model, we have evaluated the dose in two conditions: a single beam with Gaussian profile and a lateral scan. We have analysed 3 energies by evaluating the dose with the model in a 3D mesh with voxels of 1 mm dimensions. The results are obtained by selecting transverse plane at fixed depths [3].
In Fig. 3, the comparison between FLUKA and MONET profile for proton of 100 MeV at 4 cm water depth is shown. The relative error is evaluated in a central cross section and is about

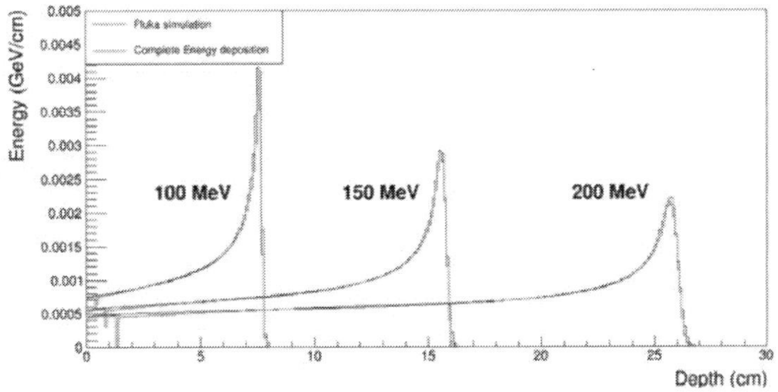

Figure 2. Bragg curves including straggling effects and nuclear contributions for $E = 100, 150$ and 200 MeV.

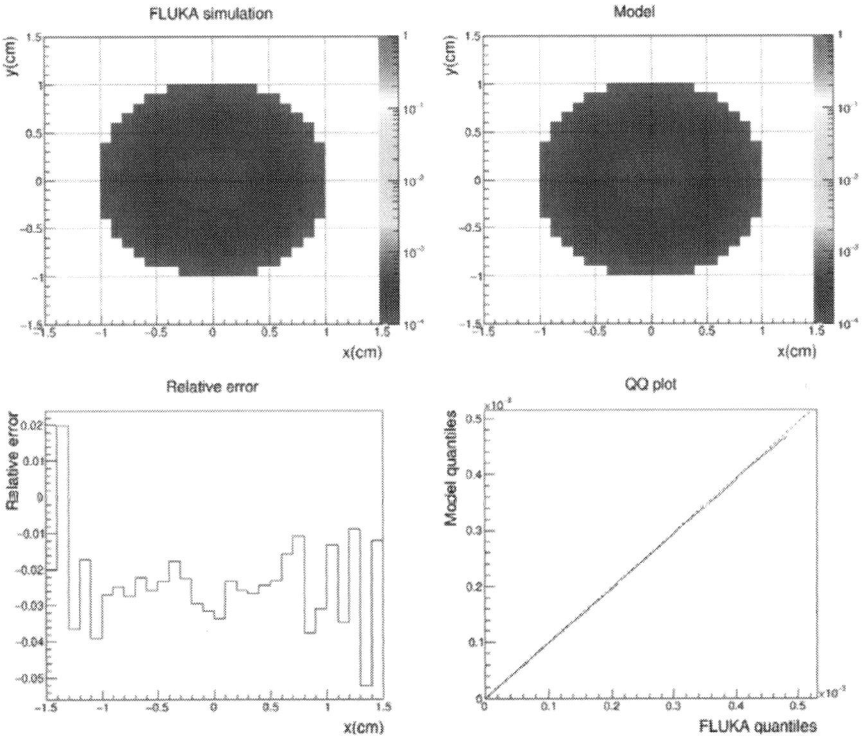

Figure 3. Energy 100 MeV at depth z=4 cm.

3%. We also present the quantile-quantile (QQ) plot to compare the quantiles of the model and MC distribution. The QQ plot is a 45° line in case of perfect match and in our case this is well verified.

In Fig.4, we have obtained a similar result for proton of 150 MeV: the agreement is good, the relative error is still 3% and also the QQ plot shows a correspondence between model and MC simulation.

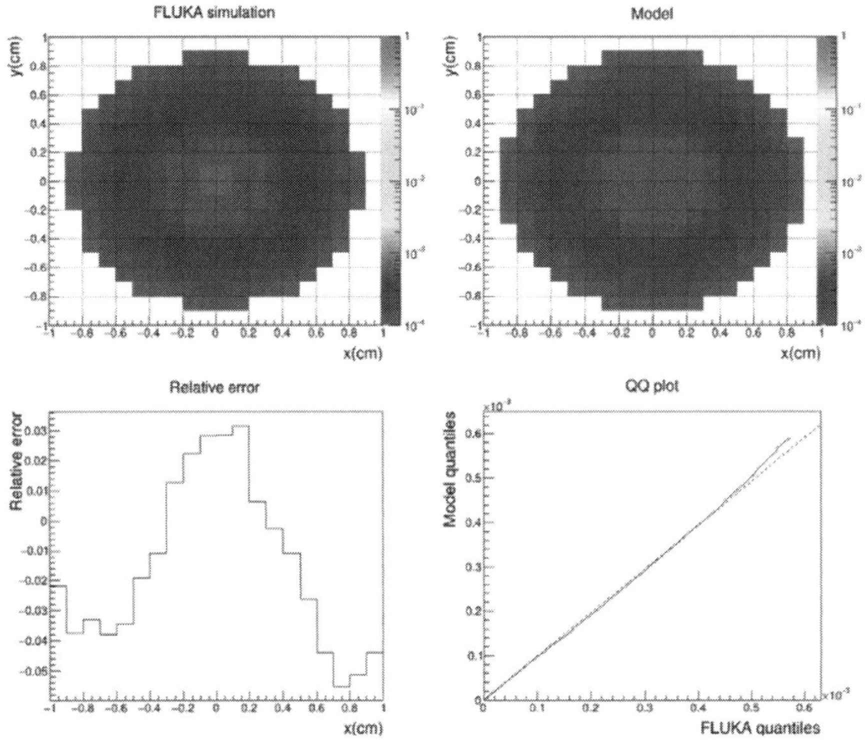

Figure 4. Energy 150 MeV at depth z=11 cm.

We have analyzed all depths for any energy and the maximum relative error in the worst case is lower than 10% in a voxel of 1 mm dimensions. It is important to note that this discrepancy once considered on the absolute energy deposition, is in a range of negligible clinical relevance. In order to estimate the accuracy of the model focusing on the tails of the distribution that give rise to the low-dose envelope, we have reproduced a lateral scan as a sum of many pencil beams. In this case, the code calculates the energy deposition for a single beam. Afterwards the dose profile is translated in order to obtain a lateral scan and the total energy deposition will be given by a sum of all single beams.
We have considered the field size factor for our model and for MC simulations. The field size factor is the ratio between the total dose deposition inside of concentric square fields of increasing size and the same quantity in a square reference D_{ref} of size 10 cm.

$$FSF = \frac{D_f}{D_{ref}}, \qquad f = 2, 4, 6, 8, 10, 12. \tag{5}$$

In Fig.5, the FSF variations as a function of square size are reported. Note that the scale in y is much expanded and the agreement between FLUKA and the model is still good.
The maximum difference is about 5%, of the same order of accuracy of the single beam tests. These results indicate a good description of the low dose contribution.

Another advantage of the MONET code is the fast calculation time. Indicatively, for each depth the calculation time is ~ 2 seconds for the single beam and ~ 4 seconds for the full lateral scan. These times are competitive compared to the simulation time ($\sim 1 - 10$ hours).

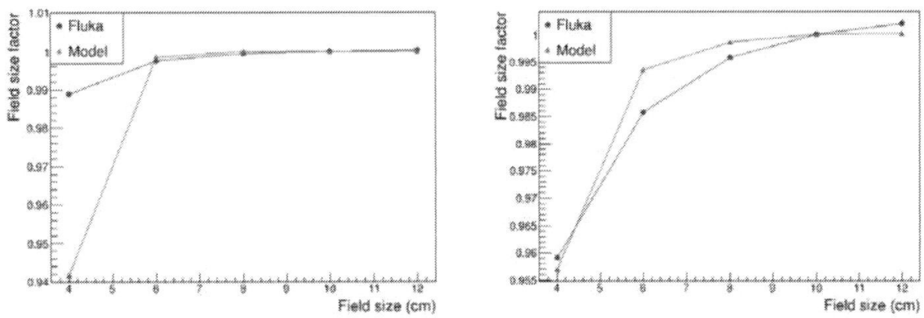

Figure 5. FSF for the model at energy 100 MeV at z=4 cm(left) and 150 MeV at z=10 cm(right). A linear interpolation of the points is added to improve visualization.

4. Conclusions

The MONET code accounts for all the physical interaction of protons with water and is based on well known and validated theories. In general, the pencil beam algorithm uses pre-calculated look-up tables for the depth dose evaluation, while the lateral profile is modelled with a Gaussian or double Gaussian distribution. On the contrary, MONET evaluates the lateral and longitudinal terms starting from first principles. The advantages of our model are the small number of parameters, the fast calculation time and the accuracy. A possible development of this study is the creation of a dose database of interest for the clinical use and an online/in-room fast dose evaluation tool.
Currently, we are extending the model to the case of Helium beam. Helium is a possible new particle for hadrontherapy: it is a good compromise between the proton and carbon ions, because it has a smaller lateral profile compared to protons and a smaller contribution of energy deposition behind the Bragg peak compared to Carbon ions.
For the lateral profile, the results of the model for Helium is in good agreement with FLUKA simulation and the accuracy is better than the parameterization just used for proton and carbon ions [14]. The last step for the implementation of MONET for ^4He beams, renamed MONETα will be the evaluation of energy deposition in the longitudinal profile.

References

[1] Loeffler J S and Durante M 2013 *Nature reviews Clinical oncology* **10(7)** 411-424
[2] Wilson R R 1946 *Radiology* **47(5)** 487-491
[3] Embriaco A *et al* 2017 *Physica Medica* **38** 6675
[4] Gottschalk B *et al* 2014 arXiv preprint arXiv:1409.1938.
[5] Parodi K *et al* 2012 *Phys. Med. Biol.* **57(12)** 3759
[6] Bellinzona V E *et al* 2016 *Phys. Med. Biol.* **61** N102
[7] Bethe H A 1953 *Physical Review* **89(6)** 1256
[8] Ulmer 2007 *Rad. Phys. and Chem.* **76** 1089-107
[9] Fruhwirth and Regler 2001 *Nucl. Instr. Meth. Phys. Res. A* **456** 369-389
[10] Papoulis 1968 *IEEE Transactions on Information Theory*
[11] Bortfeld 1997 *Med. Phys.* **24** 2024-2033
[12] Kempe *et al* 2008 *Med. Phys.* **35** 159-170
[13] Carlsson *et al* 1997 *Phys. Med. Biol.* **42** 1033-1053
[14] Embriaco A *et al* 2017 *Physica Medica* **40** 51-58

IOP Conf. Series: Journal of Physics: Conf. Series **956** (2017) 012008 doi:10.1088/1742-6596/956/1/012008

Sensitivity optimization in whispering gallery mode optical cylindrical biosensors

F Khozeymeh[1] **and M Razaghi**[2]

[1]Dept. of Physics, School of Science, University of Kurdistan, Sanandaj, Iran
[2]Dept. of Electrical Eng., School of Engineering, University of Kurdistan, Sanandaj, Iran

E-mail: fkhozeymeh@yahoo.com

Abstract. Whispering-gallery-mode resonances propagated in cylindrical resonators have two angular and radial orders of l and i. In this work, the higher radial order whispering-gallery-mode resonances, $(i = 1-4)$, at a fixed l are examined. The sensitivity of theses resonances is analysed as a function of the structural parameters of the cylindrical resonator like different radii and refractive index of composed material of the resonator. A practical application where cylindrical resonators are used for the measurement of glucose concentration in water is presented as a biosensor demonstrator. We calculate the wavelength shifts of the WG1-4, in several glucose/water solutions, with concentrations spanning from 0.0% to 9.0.% (weight/weight). Improved sensitivity can be achieved using multi-WGM cylindrical resonators with radius of $R = 100\ \mu$m and resonator composed material of MgF_2 with refractive index of $n_c = 1.38$. Also the effect of polarization on sensitivity is considered for all four WGMs. The best sensitivity of 83.07 nm/RIU for the fourth WGM with transverse magnetic polarization, is reported. These results propose optimized parameters aimed to fast designing of cylindrical resonators as optical biosensors, where both the sensitivity and the geometries can be optimized.

1. Introduction

An outstanding group of dielectric optical resonators are circular resonators which can support Whispering-gallery-mode (WGM), resonances. WGM resonators can have different geometries such as cylinder [1, 2, 3], microsphere [4, 5, 6], microfiber coils and microdisk [7]. In these geometries, due to light recycling nature, light-matter interaction is greatly enhanced. This allows fabricating of high-quality factor optical resonators, which enables confining significant optical powers in small regions for extended time periods [6]. Cylindrical micro-resonators have especially attracted a significant level of interest in biosensing [1, 2, 3]. In this work, we propose the use of cylindrical micro-resonators with optimized structural parameters for optical biosensing. This kind of optical biosensor detects changes in the surrounding refractive index through the electromagnetic (EM) field evanescent depth of WGMs, travelling outside the cylindrical resonator boundary, providing label-free sensing ability [8, 9, 10]. WGM resonances in these structures can have two degrees or orders of freedom i and l or ν, where i is radial order and l or ν is the angular order of a WGM. In this work, we analyse the high angular WGMs with radial orders $i = 1-4$, or just WG1-4, in two different polarizations transverse electric (TE), and transverse magnetic (TM). The full theoretical examination of WGMs in cylindrical resonators within the framework of classical electrodynamics requires the application of Mie theory [11]. In this paper we use an analytical model of Airy approximations to the full Mie/Debye solutions, for extracting closed form of WGM resonance wavelengths. Resonance wavelength shifts of WGMs are simulated in different concentrations of glucose-water solutions in surrounding medium of the cylindrical resonator. The sensitivity of these resonances is analysed as a function of the structural parameters of the cylindrical resonator, like different radii and different refractive indexes of composed material.

2. Theory

As mentioned, the full theoretical examination of WGMs in cylindrical resonators within the framework of classical electrodynamics requires the application of Mie theory [11, 12]. This also involves complicated computational effort due to the need for numerical calculation of the sums of series of Bessel functions [12]. This can be avoided with simpler models based on elementary optics [6] or on analytical approximations to the exact Mie/Debye solutions [13]. In the following, we will use an analytical model based on the formulation of Airy approximations to the full Mie/Debye solutions. With this formulation the calculation of WGM resonance wavelengths with different radial and angular orders is possible.

2.1. Formulation

The analytical closed form of WGM resonance wavelengths for two different polarizations of TE and TM, are expressed as follow:

$$\lambda^i_{TE}(i,l,R,m) = 2\pi n_c R(A\nu + B^i\nu^{\frac{1}{3}} - \frac{m}{\sqrt{m^2-1}} + C^i\nu^{\frac{-1}{3}} - \frac{D^i_{TE}m^3}{(m^2-1)^{\frac{3}{2}}}\nu^{\frac{-2}{3}} + o(\nu^{-1}))^{-1} \quad (1)$$

$$\lambda^i_{TM}(i,l,R,m) = 2\pi n_c R(A\nu + B^i\nu^{\frac{1}{3}} - \frac{1}{m\sqrt{m^2-1}} + C^i\nu^{\frac{-1}{3}} - \frac{D^i_{TM}(m^4-\frac{2}{3})}{m^3(m^2-1)^{\frac{3}{2}}}\nu^{\frac{-2}{3}} + o(\nu^{-1}))^{-1} \quad (2)$$

Where $\nu = l+\frac{1}{2}$, (ν and l are angular orders in spherical and cylindrical resonators respectively), $m = \frac{n_c}{n_s}$ is the relative refractive index between the cylinder (n_c), and the surrounding medium (n_s), and i is the ith zero of the Airy function and represent radial order of WGMs. With exact calculation of the numerical coefficients of A_i , B_i, C_i and D_i, WGM resonance wavelengths can be easily calculated. The superiority of using these formulations with only an error order of ν^{-1} from the exact solutions, is that these equations, provide analytical functions that can be simply implemented into a fitting routine for simultaneous determination of the parameters ν, m, and R, while the exact solution involves a difficult numerical procedure.

2.2. Equation coefficients

The exact amounts and variations of different coefficients in the above relations have been plotted in Fig. 1a-1d. It must be noted that coefficient A_i is equal 1. The numerical amount of B_i and C_i coefficients, is the same for two different polarization of TE and TM. But the D_i coefficient is different for different polarizations. These are just some mathematical results because of the airy approximation applying to the Mie/Debye solutions. With known amounts of these coefficients, sensitivity of different WGM resonance wavelengths as sensing signals can be calculated.

3. Sensitivity simulations

The operational sensing principle in WGM biosensors is to monitor changes in the WGM resonance wavelength prompted by surrounding medium refractive index changes. This phenomenon can be realized as bulk refractive index sensitivity or only sensitivity (S), of an optical biosensor. The S parameter can be expressed as below:

$$S = \frac{\triangle \lambda_{WGM}}{\triangle n_s} \tag{3}$$

where $\triangle\lambda_{WGM}$ is WGM resonance wavelength shift, over the refractive index changes of the surrounding medium, $\triangle n_s$. Indeed, interaction of the EM field evanescent depth of a WGM resonance wavelength with biological solution,(flowing in the surrounding medium), causes changes in refractive index of the surrounding medium.

The silica cylindrical resonators, with refractive index of $n_c = 1.45$ and different radii of $50 - 100\ \mu m$ are considered as optical biosensors. These parameters not only are complementary with real cylindrical device[14], but also support the propagation of multi WGM resonance. According to size parameter $x = \frac{2\pi R}{\lambda} = \frac{l}{n}$ [6], WGMs with $l > 360$, will be investigated. In the following sections of this paper, we simulate the sensitivity of the proposed cylindrical sensor in several glucose/water solutions in surrounding medium, with concentrations spanning from 0.0% to 9.0%. The change in refractive index of the solution $\triangle n_s$, is proportional to the glucose concentration C through the relation $\triangle n_s = C \times 1.375 \times 10^{-3} RIU/\%$ as reported in

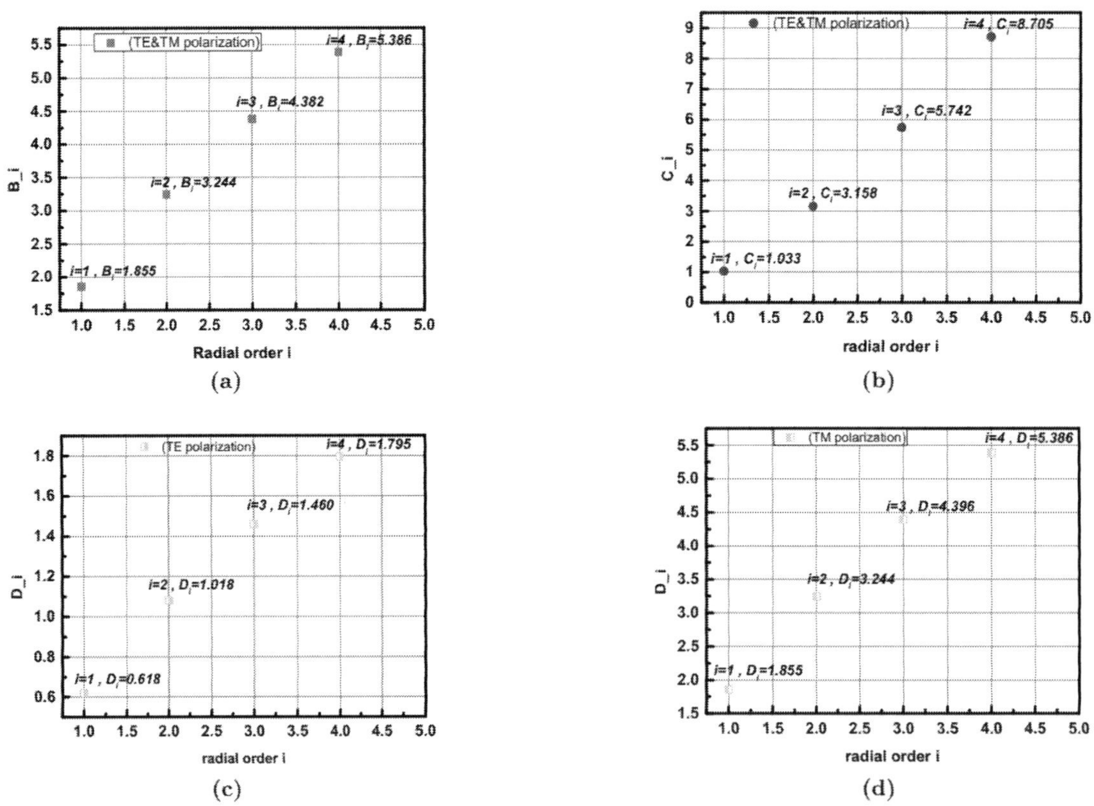

Figure 1. Variations of coefficients of a) B_i and b) C_i for both polarization TE and TM and c) D_i for TE and d) D_i for TM polarization, in terms of radial order $i = 1 - 4$

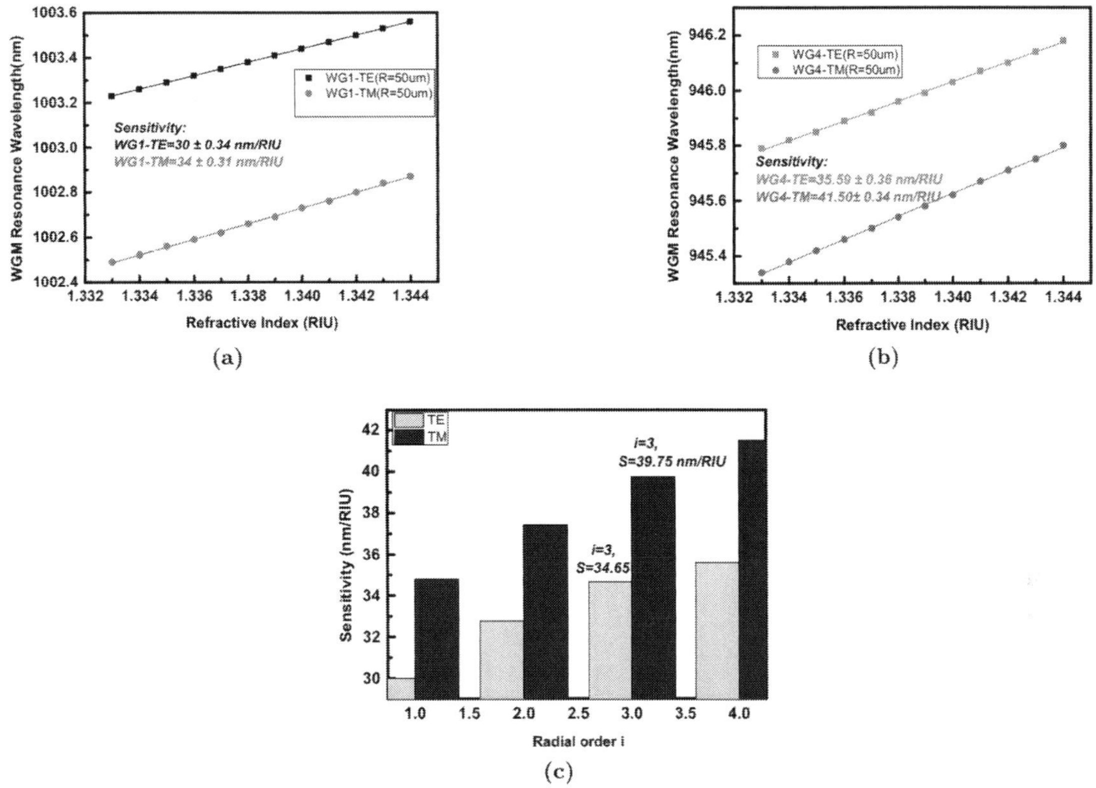

Figure 2. WGM resonance wavelength shift in different refractive index of glucose-water for TE and TM a) WG1 at $R = 50$ μm , b) WG4 at $R = 50$ μm and c) sensitivity of WG1-4 at fixed $R = 50$ μm.

[15]. WG1-4, in a fixed angular order $l = 442$ and their effective role in sensing performance, are studied. We have especially examined WGMs with $l = 442$, because the WGMs with this angular momentum have experimentally demonstrated in [14] as sensing signals in optical cylindrical biosensors. Also, we considered different composed material of cylindrical resonators to optimize the structural parameters for better sensing performance. It has a great significance for providing a theoretical guide for the fabrication of the device and its practical sensing applications.

3.1. Radial order and polarization
Each WGM has its unique S. Fig. 2a-2c shows the effect of both radial order and polarization on sensitivity. Higher radial order WGMs, have a longer evanescent depth of electromagnetic field into surrounding medium. This is why they have higher sensitivity. Also polarization change, has an effective role on S of WGM resonance wavelengths.

3.2. Resonator radius
Then we investigate the effect of different radii $R = 50 - 100$ μm on sensitivity. Fig. 3a-3c shows the calculated sensitivity versus R under different radial orders $i = 1 - 4$ and polarizations TE and TM, while $n_c = 1.45$ remains unchanged. Higher radial order WGMs show more sensitivity. As they have more EM evanescent depth in surrounding medium. Also resonator radius increasing has a positive effect on sensitivity improvement. As we can see Fig. 3b, with

increasing resonator radius from $R = 50$ μm to $R = 100$ μm, S has increased from 35 nm/RIU to 71 nm/RIU.

3.3. Composed materials of the resonator

Calculated sensitivity of our device with different WGMs versus the resonator refractive index n_c has been plotted in Fig. 4a-4b. It can be seen that lower resonator refractive index possesses significantly higher sensitivity.

4. Conclusions

We proposed optimized structures of optical biosensors based on multi-WGM cylindrical resonators formed in silica glass with refractive index of $n_c = 1.45$. Analytical closed form of WGM resonance wavelengths with radial orders $i = 1 - 4$ have been calculated. Exact amounts of equation coefficients in these relations, have been extracted. Results showed a bulk refractive index sensitivity of $\sim$ 71 nm/RIU for TE WG4 and $\sim$ 83 nm/RIU for TM WG4 at resonator radius of R=100 μm. We indicated that higher radial orders of WGMs at bigger radii of resonator, and lower refractive index of the resonator material like MgF_2 material, lead to higher sensitivity. These results will be the theoretical guides for fast designing and fabrication of the optical biosensor device, where both the sensitivity and the geometries can be optimized.

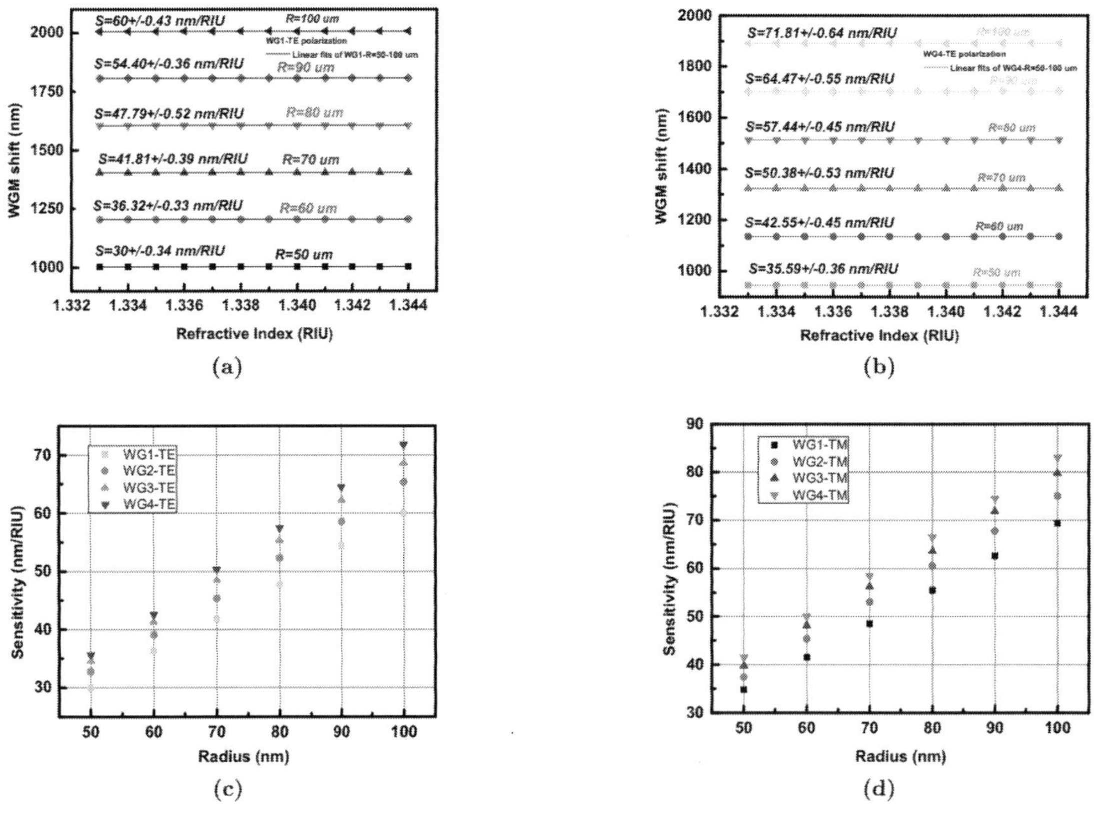

Figure 3. a) and b) resonance wavelength shifts of WG1 and WG4 in different surrounding medium refractive index for TE polarization in different radii, $R = 50 - 100$ μm of resonator; c) and d) biosensor sensitivity in different radii of cylindrical resonator for TE WG1-4 and TM WG1-4, respectively.

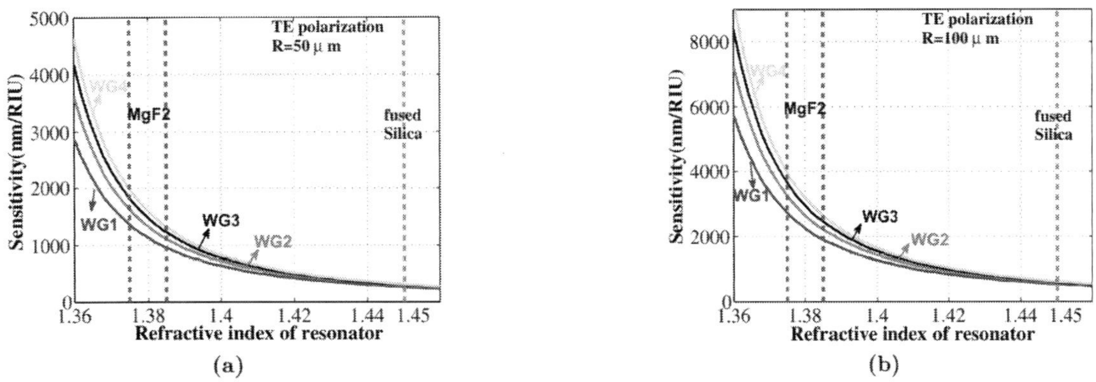

Figure 4. Sensitivity of WG1-4 as a function of resonator refractive index in resonator radius of a) $R = 50$ μm and b) $R = 100$ μm.

References

[1] Jin X, Dong Y and Wang K 2015 *Applied optics* **54** 8100–8107
[2] Jin X, Dong Y, Wang K and Jian H 2016 *IEEE Photonics Technology Letters* **28** 1649–1652
[3] Sumetsky M 2010 *Optics letters* **35** 2385–2387
[4] Chiasera A, Dumeige Y, Feron P, Ferrari M, Jestin Y, Nunzi Conti G, Pelli S, Soria S and Righini G C 2010 *Laser & Photonics Reviews* **4** 457–482
[5] Francois A and Himmelhaus M 2009 *Sensors* **9** 6836–6852
[6] Righini G, Dumeige Y, Féron P, Ferrari M, Nunzi Conti G, Ristic D and Soria S 2011 *Rivista del Nuovo Cimento* **34** 435–488
[7] Schweinsberg A, Hocdé S, Lepeshkin N N, Boyd R W, Chase C and Fajardo J E 2007 *Sensors and Actuators B: Chemical* **123** 727–732
[8] Guider R, Gandolfi D, Chalyan T, Pasquardini L, Samusenko A, Pederzolli C, Pucker G and Pavesi L 2015 *Sensing and Bio-Sensing Research* **6** 99–102
[9] Guider R, Gandolfi D, Chalyan T, Pasquardini L, Samusenko A, Pucker G, Pederzolli C and Pavesi L 2015 *Sensors* **15** 17300–17312
[10] Chrostowski L, Grist S M, Flueckiger J, Shi W, Wang X, Ouellet E, Yun H, Webb M, Nie B, Liang Z *et al.* 2012 *Silicon photonic resonator sensors and devices.* Ph.D. thesis University of British Columbia
[11] Bohren C F and Huffman D R 2008 *Absorption and scattering of light by small particles* (John Wiley & Sons)
[12] Roll G, Kaiser T and Schweiger G 1998 *Applied optics* **37** 2483–2492
[13] Lam C, Leung P T and Young K 1992 *JOSA B* **9** 1585–1592
[14] Fan X, White I M, Zhu H, Suter J D and Oveys H 2007 *Proc. SPIE* vol 6452 p 64520M
[15] bin Mat Yunus W M and bin Abdul Rahman A 1988 *Applied optics* **27** 3341–3343

IOP Conf. Series: Journal of Physics: Conf. Series **956** (2017) 012009 doi:10.1088/1742-6596/956/1/012009

Spectroscopic techniques to study the immune response in human saliva

E Nepomnyashchaya, E Savchenko, E Velichko, T Bogomaz and E Aksenov

Institute of Physics, Nanotechnology and Telecommunications, Peter the Great Saint Petersburg Polytechnic University, Saint Petersburg 195251, Russia

E-mail: elina.nep@gmail.com

Abstract. Studies of the immune response dynamics by means of spectroscopic techniques, i.e., laser correlation spectroscopy and fluorescence spectroscopy, are described. The laser correlation spectroscopy is aimed at measuring sizes of particles in biological fluids. The fluorescence spectroscopy allows studying of the conformational and other structural changings in immune complex. We have developed a new scheme of a laser correlation spectrometer and an original signal processing algorithm. We have suggested a new fluorescence detection scheme based on a prism and an integrating pin diode. The developed system based on the spectroscopic techniques allows studies of complex process in human saliva and opens some prospects for an individual treatment of immune diseases.

1. Introduction

Antibacterial properties of a human body play an important role in protection of an organism against an infection [1]. The ability to kill microbes is an important parameter in assessment of the immune status of the person especially at pathology [2]. Consequently, it is important to estimate activity of humoral and cell mediated immune responses. The humoral immunity works through various substances like blood plasma, saliva and glandula secretions which are capable to suppress reproduction of microbes [3]. Among these substances, antibodies or immunoglobulins play one of the most important roles in immune response [4]. Antibodies react to specific antigens [1] — the alien substances which find their way into a body.

The immune response in an organism is caused by several of factors. So, to characterize the immune response one should investigate concentrations of these or those factors of immunity [2]. To select the effective personalized treatment, it is necessary to estimate also their activity, i.e. ability of antibodies to bind to antigens of the infectious agents which got to an organism. In modern medicine different optical methods are expected to be promising for therapy of infections and immune diseases [5, 6, 7].

In our work we used two spectroscopic techniques, laser correlation spectroscopy and fluorescence spectroscopy, for the immune system activity investigation. In previous work it was concluded that laser correlation spectroscopy allows estimation of the ability of biomolecules in biological fluids in dynamics [8]. Dynamic parameters are essential for immune studies because the reaction time shows the general immune activity. We decided to extend laser correlation spectroscopic experiments with fluorescence spectroscopic technique to get some new information about immune system such as conformational changes in immunoglobulins and other molecules in dynamics.

2. Materials and methods

2.1 Laser correlation spectroscopy

Laser correlation spectroscopy allows one to estimate sizes of nanoparticles in a fluid from measurements of scattered radiation parameters [9]. The experiments include recording of the time dependence of the scattered light intensity and the calculation of its autocorrelation function. According to the dynamic scattering theory [9], the autocorrelation function of the light scattered by particles in a solution will carry information on the diameters (D) of these particles.

To derive information about sizes of nanoparticles the distribution of scattered light intensity I(D) is counted. An algorithm based on Tikhonov regularization method and described in [10] is used for calculation of intensity distribution I(D) from autocorrelation function.

A typical scheme of the laser correlation spectroscopy device is shown in Fig. 1. In this setup, a laser beam is transmitted through a converging lens and focused on the cell containing the sample. Laser light is recorded at the angle $\theta = 90°$ and passes through single mode optical fiber to a photomultiplier. The signal from the photomultiplier is digitized and arrives to a computer for the correlation analysis. More specific description can be found in [10, 11].

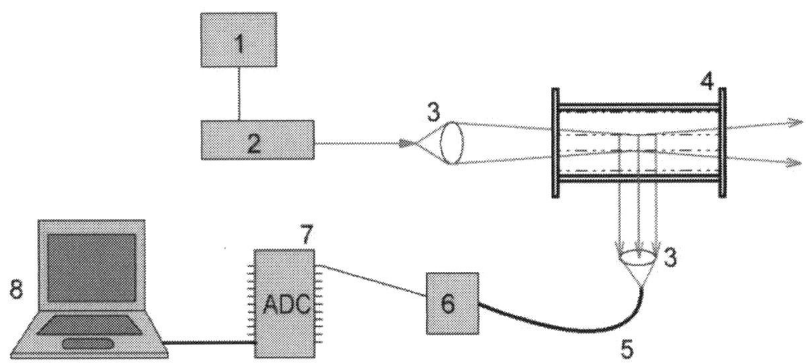

Figure 1. Scheme of a laser correlation spectrometer. 1 — power supply for laser; 2 — laser radiation source; 3 — converging lens; 4 — sample; 5 — optical fiber; 6 — photomultiplier; 7 — analog-to-digital convertor; 8 — computer.

2.2 Fluorescence spectroscopy

In fluorescence spectroscopy the fluorescence spectra and intensities are usually registered [12]. By measuring changes in the fluorescence intensity important conclusions about immune processes can be made, such as the general composition of antibodies, their conformational structure and activity [12]. A scheme of the experimental setup is shown in Fig. 2. Light from UV laser module passes through the attenuating filters, interacts with the sample deposited on the prism bottom causing its fluorescence. The fluorescence signal is detected by using a dynamic pin-diode [13], placed directly under the sample. Due to this design with a total internal reflection prism, the main beam does not pass to the pin diode and only the fluorescence signal intensity is measured. The output signal is analyzed on computer to calculate the fluorescence intensity [13].

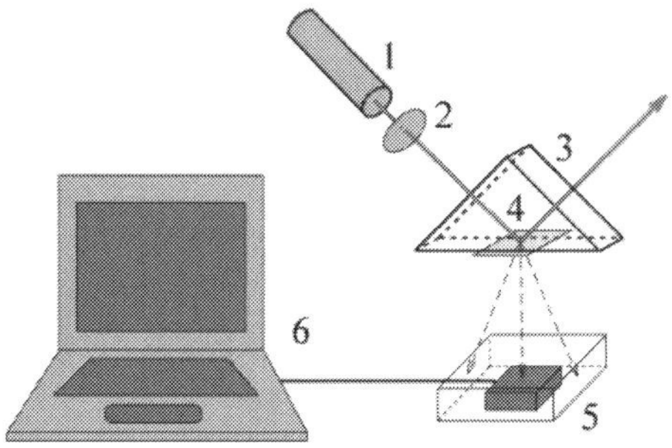

Figure 2. Experimental setup: 1 — laser, 2 — attenuating filter, 3 — total internal reflection prism, 4 — sample, 5 — pin-diode, 6 — computer.

2.3 Object of study

In this work mixed saliva (from all salivary glands) was chosen as the test biological solution. This biological material performs varied functions (cleansing, protective, immune, hormonal), because of its complex biochemical composition. It consists of proteins, lipids, carbohydrates, ions of sodium, potassium, vitamins et al. [14, 15]. Therefore, mixed saliva can be used for immune activity analysis. Promising new technologies have unveiled large numbers of medically valuable salivary biomarkers for different disease conditions including cancer, autoimmune, viral, bacterial, cardiovascular, and metabolic diseases [15 – 18]. Due to noninvasiveness of sample collection, smaller sample aliquots, cost effectiveness, easy storage and transportation, greater sensitivity, and correlation with levels in blood [19] mixed saliva has a great perspective in modern medical diagnostics [20, 21].

In our experiments the mixed saliva solutions with fluorescence spectrum in the range from 300 nm to 400 nm were used. A vaccine of tick-borne encephalitis containing antigens was used for triggering of immune response processes in saliva solutions. It was assumed that mixing saliva solutions with tick-borne encephalitis would provide binding of immunoglobulins with antigens in the vaccine.

Used chemical agents:

1. Aluminium hydroxide buffer;
2. Veronal buffer (VBS);
3. Natural saline solution;
4. Dried tick-borne encephalitis vaccine;
5. Saliva taken from healthy patient (women)

The mixed saliva was dissolved in VBS in 25% concentration. The tick-borne encephalitis vaccine was dissolved in aluminium hydroxide buffer with addition of natural saline solution.

3. Results and discussion

In the fluorescence measurements 0.02 ml of sample was placed on the glass under the prism. The fluorescence intensity was measured immediately after the samples preparation, after 45 minutes of incubation under the room temperature and after 2 hours of incubation. The measured fluorescence intensities are summarized in Tab. 1.

Table 1. Calculated fluorescence intensities for mixed saliva with vaccine of tick-borne encephalitis addition

Object	Immediately after preparation	45 min after preparation	2 hours after preparation
Saliva dissolved in VBS with vaccine of tick-borne encephalitis	14.29 mW/mm^2	1.06 mW/mm^2	0.45 mW/mm^2

It can be noticed that the fluorescence intensity of saliva after addition of vaccine decreased while incubation. The decrease in fluorescence intensity can indicate reorganization of tryptophan – the fluorescent amino acid, part of almost all immune proteins [22]. The tryptophan fluorescence is higher when it is not blocked by other proteins parts. It was concluded that the fluorescence decrease suggests possible conformational changes in the structure of immune complexes and antigen antibody reactions. Saliva solution reacts with an antigen (vaccine of tick-borne encephalitis), blocking the tryptophan fluorescence, therefore the total fluorescence intensity decreases [22].

To confirm the existence of immune system reaction, measurements of particle sizes in saliva solution before and after the vaccine addition were taken. The results are presented in Fig. 3 – 6.

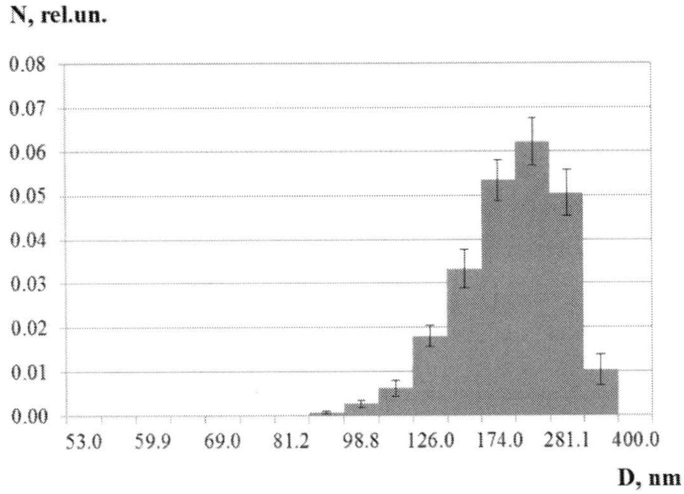

Figure 3. Sizes of particles in saliva dissolved in VBS: 45 min after preparation.

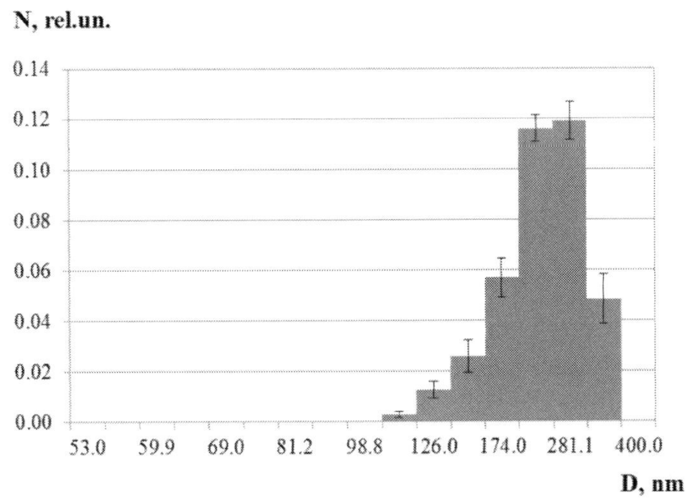

Figure 4. Sizes of particles in saliva dissolved in VBS: 120 min after preparation.

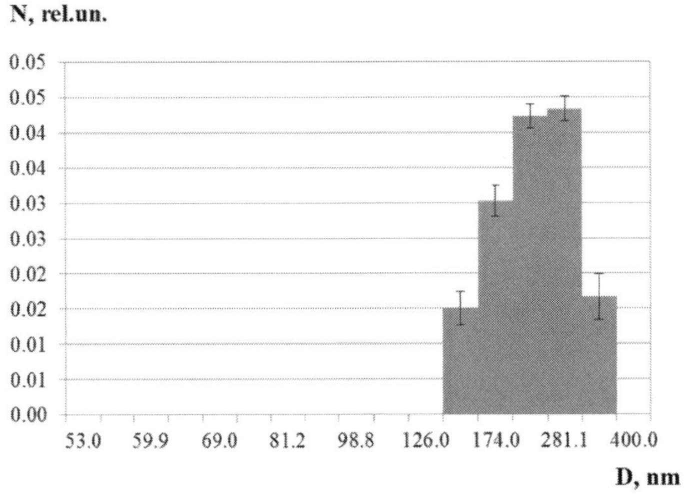

Figure 5. Sizes of particles in saliva dissolved in VBS with vaccine of tick-borne encephalitis addition: 45 min after preparation.

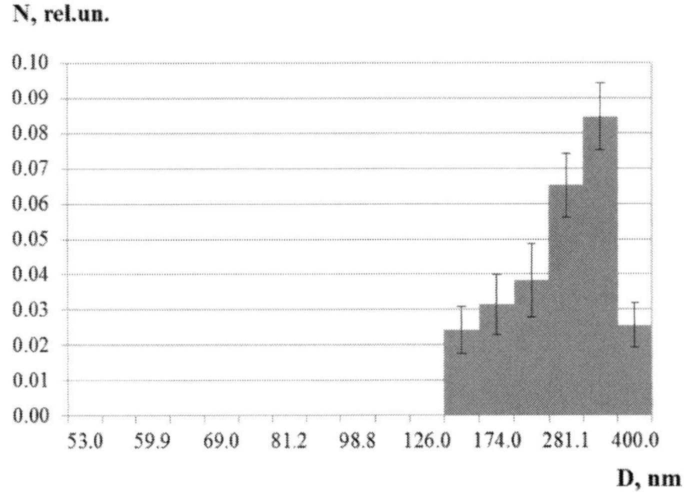

Figure 6. Sizes of particles in saliva dissolved in VBS with vaccine of tick-borne encephalitis addition: 120 min after preparation.

It can be observed that in the saliva solution there are no noticeable changes in particle sizes while incubation within 45 minutes – 2 hours. This demonstrates lack of reactions in the studied solution. However after addition of vaccine of tick-borne encephalitis (the measured particle size in pure vaccine equals to 5.8 ± 0.4 nm) in solution of saliva the sizes of particles are increased noticeably (minimal sizes increased from 100 nm to 150 nm) after 45 minutes of incubation and further incense (maximal sizes increased from 300 nm to 400 nm) after 2 hours of incubation. These results clearly demonstrate reaction of humoral immunity in human saliva on the antigen. By measuring sizes we could detect the degree of aggregation of immune complexes with antigens in vaccine: higher sizes, stronger reaction. In addition, we could measure time of immune reaction that shows how fast an organism can detect and incapacitate infections.

4. Conclusion

In this work the possibilities of spectroscopic techniques for medical diagnostics on the example of immune studies are demonstrated. By using the fluorescence spectroscopic technique we detected some conformational changes in saliva molecules showed by tryptophan fluorescence decreasing when the vaccine of tick-borne encephalitis was added. We suggested that the decrease in tryptophan fluorescence is a result of its blocking in the center of molecules and also about its aggregation. Laser correlation spectroscopic technique confirmed the aggregation theory and showed increasing aggregation sizes in time. By determining sizes and time of aggregation exploiting fluorescence measurements the immune response for each individual person could be measured. The further studies of immune system in human saliva and also in blood could open new perspectives for cheap diagnostics and individual therapy.

Acknowledgements

Authors are very grateful ActLight SA (Lausanne, Switzerland, http://act-light.com/technology) for their donation of pin-diode for that research. Also authors would like to thank O Tsybin and D Dyubo for great help in providing of fluorescence experiments.

References

[1] Abbas A, Lichtman H and Pillai S 2014 *Els. Health Sc.* p. 544
[2] Brodin P and Mark D 2017 *Nat. Rev. Immunol.* **17** 21–29
[3] Gibbins H, Proctor G, Yakubov G, Wilson S and Carpenter G 2014 **20** *Oral des.* 707–713
[4] Schwab I and Nimmerjahn F 2013 *Nat. Rev. Im.* **13** 176-189
[5] Hemphill J, Borchardt E, Brown K and Asokan A 2015 *Am. Chem. Soc.* **137** 5642–5645
[6] Tan P, He L, Han G and Zhou Y 2017 *Tr. in Biotech.* **35** 215–226
[7] Weissleder R and Nahrendorf M 2015 *Proc. Natl. Acad. Sci. USA* **112** 14424–14428
[8] Stetefeld J, McKenna S and Patel T 2016 *Biophys. Rev.* **8** 409–427
[9] Nepomnyashchaya E K, Prokofiev A V, Velichko E N, Pleshakov I V and Kuzmin Yu I 2017 *J. MMM* **431** 24–26
[10] Nepomnyashchaya E, Velichko E and Aksenov E 2016 *IOP: Conf. Ser.* **769** 012025
[11] Nepomnyashchaya E, Savchenko E, Velichko E and Aksenov E 2017 *Proc. SPIE* **10336** 103360H
[12] Roederer M, Quaye L, Mangino M, Beddall M and Mahnke Y 2015 *Cell* **161** 387–403
[13] Okhonin S, Gureev M, Salin D, Appel J, Kvasov A, Pastre M, Polzik E S, Tagantsev A K, Uddeegard F and Kayal M 2015 *J.App. Ph. Lett.* **06** 3111
[14] Yuavaraj M, Udayakumar K, Jayanth V, Rao A, Bharanidharan G, Koteeswaran D, Munusamy B, Muarali Krishna C and Ganesan S 2014 *J. of Photochem. and Photob. B:Biol.* **130** 153-16
[15] Lamont R, Hajishengallis N and Jenkinson F 2014 *Oral microb. and imm.* 504
[16] Pfaffe T, Cooper-White J, Beyerlein P, Kostner K and Punyadeera C 2011 *Clin. Chem.* **57** 675–687
[17] Slavish D, Graham-Engeland J, Smyth J and Engeland C 2015 *Brain, Beh. and Imm.* **44** 253–269
[18] Escribano D, Gutiérrez A, Tecles F and Cerón J 2015 *Res. in Vet. Sci.* **102** 38–44
[19] Malathi N, Mythili S and Vasanthi H 2014 *ISRN Dent* **2014** 158786
[20] Yakob M, Fuentes L, Wang M, Abemayor A and Wong D 2014 *Cur. Oral Health Rep.* **1** 133–141
[21] Korte D and Kinney J 2016 *Period.* **70** 26–37
[22] Ghisaidoobe A and Chung S 2014 *Int. J. Mol. Sci.* **15** 22518–22538

IOP Conf. Series: Journal of Physics: Conf. Series **956** (2017) 012010 doi:10.1088/1742-6596/956/1/012010

Investigation of some possible changes in Am-Be neutron source configuration in order to increase the thermal neutron flux using Monte Carlo code

H Basiri*[1], H Tavakoli-Anbaran[1]

[1] Faculty of Physics and Nuclear Engineering, Shahrood University of Technology, Shahrood, Semnan, Iran

E-mail: basiri@mail.um.ac.ir

Abstract. Am-Be neutrons source is based on (α, n) reaction and generates neutrons in the energy range of 0-11 MeV. Since the thermal neutrons are widely used in different fields, in this work, we investigate how to improve the source configuration in order to increase the thermal flux. These suggested changes include a spherical moderator instead of common cylindrical geometry, a reflector layer and an appropriate materials selection in order to achieve the maximum thermal flux. All calculations were done by using MCNP[1] Monte Carlo code. Our final results indicated that a spherical paraffin moderator, a layer of beryllium as a reflector can efficiently increase the thermal neutron flux of Am-Be source.

1. Introduction

The neutrons are unstable particles that with the protons form the atomic nuclei. Free neutrons can be produced by the cosmic rays showers but also dedicated neutron sources as neutron generators, research reactors and spallation sources are used to produce them artificially [1]. Nuclear reactors are known as the best source of neutrons and can supply high neutron fluxes, but the high technology involved, the safety and security concerns, the costs and their huge dimensions make it impossible to use reactors in many applications [2]. (α, n) sources are a good alternative for reactors. In this kind of source a heavy nucleus emits alpha particles, absorbed by another nucleus that will produce a neutrons. The most common neutrons emitter is the beryllium, which decays in the following way [3]:

$$\alpha + {}^{9}Be \rightarrow {}^{12}C + n$$

Am-Be source yields $2.2 \times (10)^6$ neutrons per second per curie. Its half-life is about 458 years and has low gamma emission rate. Depending on the different application, the proper energy range of the neutrons can be chosen. In this study, we focused on thermal neutrons and the goal is to increase thermal flux. Fig.1 indicates the geometry of the neutron source used in our calculations. The source is fabricated with a compacted mixture of americium oxide (AmO_2) beryllium metal powder doubly encapsulated in welded stainless steel [4].

[1] Monte Carlo N-Particle

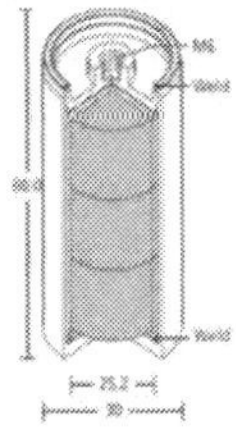

Figure 1: Encapsulated Am-Be source. (X.14 model)

1.1. Neutron moderators
As mentioned above, Am-Be source produces neutrons in wide energy range. In order to thermalize the fast neutrons a moderator is needed. An appropriate moderator is expected to have a very low thermal neutron cross section and a low atomic number in order to increase the neutron energy loss on each collision. Some suggested materials are C, H_2O, D_2O, Be, and Polyethylene [5].

1.2. Reflectors
In reactors a reflector material is used to return the neutrons that are leaking from the surface. This material can reduce the consumption of reactor's fuel. A good reflector should have low absorption and high reflection for neutrons, high resistance to oxidation and irradiation [6]. Beryllium, heavy water and graphite are some good example of reflector.

2. Simulations
We used MCNP Monte Carlo code for all next simulations [7]. Results in this code are averaged and reported for one neutron. In the first step, the effect of moderator geometry on the thermal flux was investigated. Then we used different moderators to find which one is most suitable for our purpose. In the next step, the appropriate radius of the selected moderator was determined. In the final section, we investigate the effect of adding reflectors to the source configuration. The statistical error in simulations, caused by using Monte Carlo method, was less than 10% and the effect of chemically bound atoms $[S(\alpha, \beta)]$ on the thermal neutron scattering is not considered.

2.1. Effect of moderator geometry on the neutron flux
According to the simple diffusion theories the leakage of neutrons is proportional to the surface area of the core and for a spherical core the surface area per unit of volume is a minimum. Therefore, for a spherical core, the fraction of neutrons that are lost by leakage is smaller than for any other geometry. To compare the effect of geometry on the thermal flux, spherical and cylindrical water moderator with the same volume were simulated and the thermal flux was determined using F4 tally, which calculates the average neutron flux in a volume [8]. Results are indicated in Fig. 2. Simulations showed that the spherical moderator allow to obtain a greater thermal flux with respect to the cylindrical one.

2.2. Selecting the best moderator
To compare different moderators, the thermal neutron flux in various materials with different radiuses was calculated using F2 tally, which counts the average neutron flux on a surface [8]. In Tab. 1 The evaluated thermal neutron flux on the surface of moderator sphere are reported. Results are depicted in Fig. 3 and Fig. 4 for six various type of moderators.

IOP Conf. Series: Journal of Physics: Conf. Series **956** (2017) 012010 doi:10.1088/1742-6596/956/1/012010

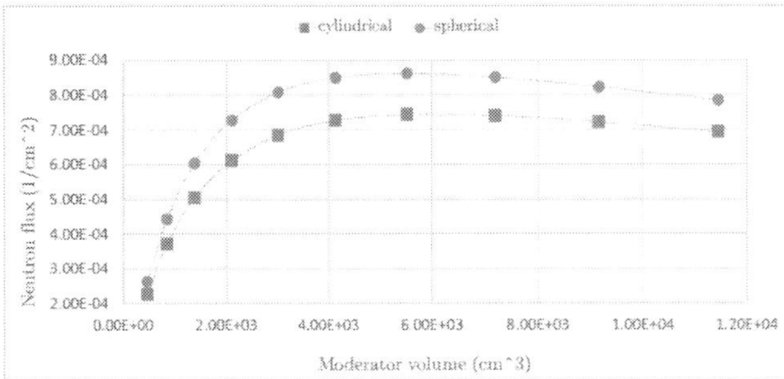

Figure 2: Comparison of the thermal neutron flux in cylindrical and spherical moderators.

Table 1. Thermal neutron flux in various materials with different radius.

Radius(cm)	BeO	D2O	H2O	Paraffin	Polyethylene	Graphite
4	6.99E-11	2.02E-09	3.58E-05	7.65E-05	6.97E-05	0
5	7.19E-10	1.24E-08	7.31E-05	1.34E-04	1.25E-04	0
6	1.23E-08	6.55E-08	1.03E-04	1.66E-04	1.58E-04	0
7	1.08E-07	2.32E-07	1.21E-04	1.74E-04	1.68E-04	1.35E-11
8	4.91E-07	6.27E-07	1.26E-04	1.66E-04	1.62E-04	3.30E-10
9	1.49E-06	1.36E-06	1.23E-04	1.49E-04	1.47E-04	2.36E-10
10	3.50E-06	2.52E-06	1.15E-04	1.29E-04	1.28E-04	1.15E-09
11	6.71E-06	4.16E-06	1.04E-04	1.09E-04	1.09E-04	3.72E-09
12	1.11E-05	6.29E-06	9.14E-05	8.99E-05	9.13E-05	1.07E-08

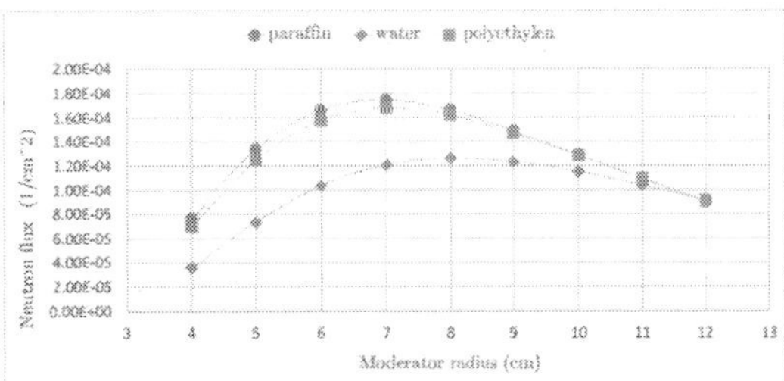

Figure 3: Neutron thermal flux vs. moderator sphere radius.

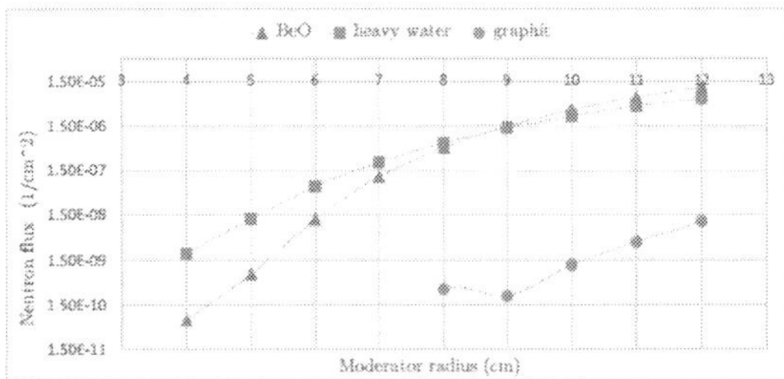

Figure 4: Neutron thermal flux vs. moderator sphere radius.

According to the results on the surface of a paraffin sphere of radius of 7 cm, the thermal flux is maximum.

2.3. Determination of the optimized radius for the moderator

In order to reduce the effect of fast neutrons in the paraffin, the radius was increased by 0.2λ, where λ is the neutron mean free path for paraffin [9]. The radius is selected in order to have the thermal flux almost equal to the non-thermal one: as reported in Fig. 5 this happens for a radius of 31.42 cm.

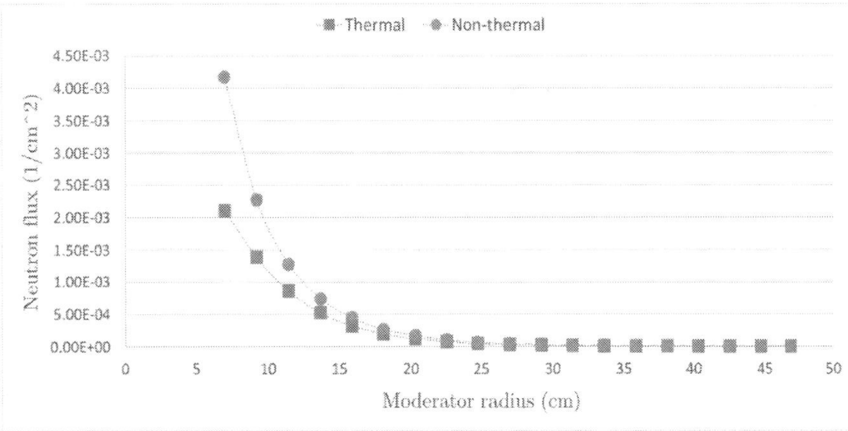

Figure 5: Comparison of fast and thermal flux of different moderator radiuses.

2.4. Reflectors

To find the most efficient reflector, the Am-Be source and its surrounding paraffin moderator were embedded in different reflectors. As a first step, the effect of adding different thicknesses of reflectors to a sphere of a paraffin of radius of 31.42 cm was investigated. Neutron flux on two detectors which were located at 7 cm and 31.42 cm was calculated. Results indicates that adding the thickness of reflectors to this radius of paraffin can not affect the thermal flux on the radius of 7 cm, but it can increase the flux on the surface of paraffin sphere of the 31.42 cm radius. The thermal flux obtained on the surface of paraffin sphere of radius 31.42 cm with different reflectors are reported as a function of thickness in Fig.6.

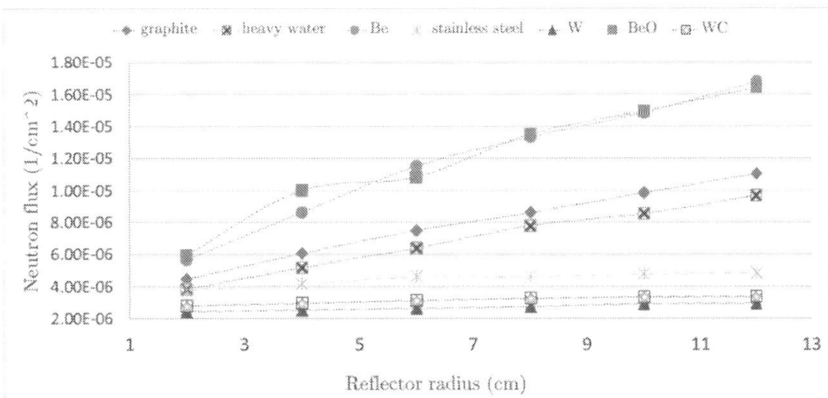

Figure 6: Neutron flux graph vs. radius.

Comparing the results, it is founded that the beryllium is the best reflector.

Since adding the reflector to a sphere of 31.42 cm of moderator cannot make any increase in the radius of 7 cm, the reflector effects in the case of a paraffin sphere of a radius of 7 cm were investigated. In this section, three reflectors were chosen and the results are depicted in Fig. 7.

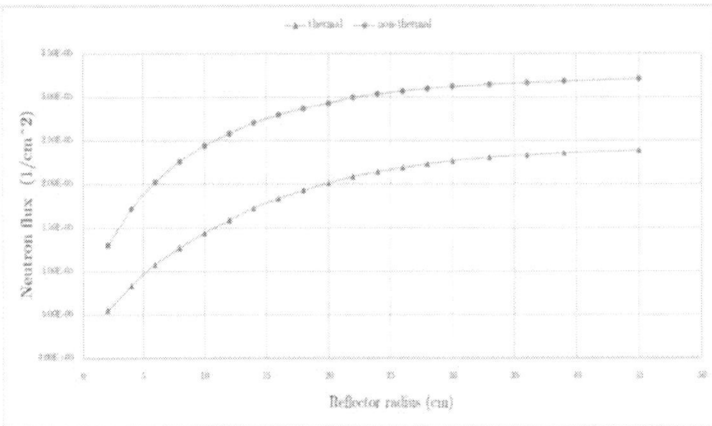

Figure 7: Effect of adding different thicknesses of three reflectors.

The simulations show that the addition of a reflector layer can lead to an increase in the flux of thermal neutrons. As we expected, beryllium had the greatest effect on the flux of thermal neutrons.

2.5. Finding the Effective radius of the reflector

To find out how the thickness of the reflector will affect the neutron fluxes, we increased the radius of the Be and calculate both thermal and non-thermal fluxes on the surface of the paraffin sphere of 7 cm. Results are reported in Fig. 8.

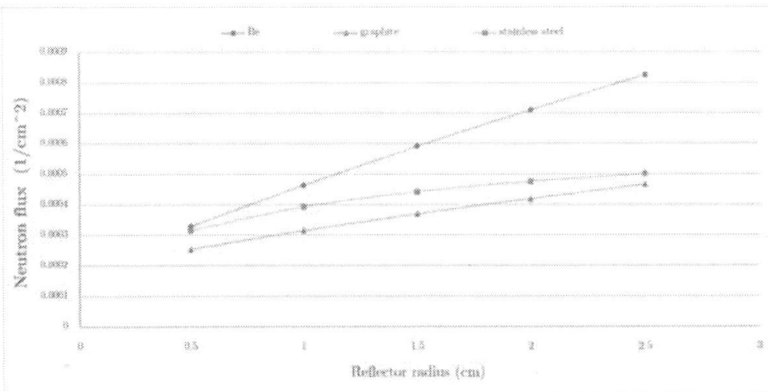

Figure 8: Effect of adding different thicknesses of Be reflector.

The results showed that the addition of beryllium initially led to an increase in the flux of thermal and non-thermal neutrons but the effect starts to be less evident for a reflector radius greater than 30 cm, while the differences among the thermal flux and the not-thermal one are quite constant. Finally a paraffin sphere of a radius 7 of cm with a 30 cm of beryllium proves to be the best solution. Fig. 9 indicates the final suggested geometry in MCNP code.

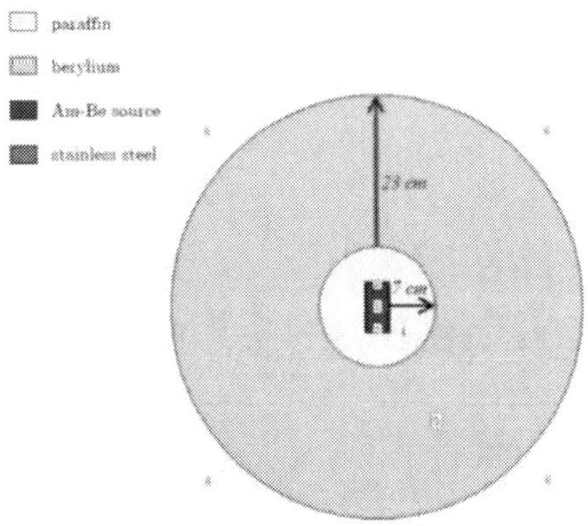

Figure 9: The final suggested source configuration in MCNP code.

3. Conclusion

The performed simulations indicated that using a spherical moderator instead of common cylindrical moderators can led to an increase in thermal neutron flux. Among the different moderators, paraffin produced the highest flux of thermal neutrons in the surface of a sphere of 7 cm radius, where we suggest to put the end of the irradiation site. Moreover, beryllium was established as the best neutron reflector to increase the thermal flux of an Am-Be neutrons source.

References

[1] Craft A E 2012 *Design, construction, and demonstration of a neutron beamline and a neutron imaging facility at a MARK-I TRIGA reactor* Ph.D. thesis Colorado School of Mines Colorado, United States
[2] Knoll G F 2010 *Radiation Detection and Measurement* (John Wiley & Sons) ISBN 978-0-470-13148-0
[3] Schieck H P g 2014 *Nuclear Reactions: An Introduction* (Springer) ISBN 978-3-642-53986-2
[4] Eckert and Company Z 2009 *Sealed Radiation Sources Product Information* (Nuclitec GmbH)
[5] Bennet D J and Thomson J R 1989 *The elements of nuclear power* (Longman Scientific & Technical) ISBN 978-0-470-21317-9
[6] Agarwal S K 2003 *Nuclear Energy: Principles, Practices, and Prospects* (APH Publishing) ISBN 978-81-7648-444-2
[7] Schwarz A and Schwarz R 2008 *MCNP/MCNPX Visual Editor Computer Code Manual*
[8] Briesmeister J 1985 *MCNP: A General Monte Carlo Code for Neutron and Photon Transport* (Los Alamos National Laboratory)
[9] Aygn B and Budak G 2012 *Nuclear Science and Technology* 33–39

IOP Conf. Series: Journal of Physics: Conf. Series **956** (2017) 012011 doi:10.1088/1742-6596/956/1/012011

The CUORE experiment at LNGS

V Dompè[1,2] on behalf of CUORE Collaboration

[1]INFN – Gran Sasso Science Institute, L'Aquila I-67100, Italy

[2]INFN – Laboratori Nazionali del Gran Sasso, Assergi (L'Aquila) I-67100, Italy

E-mail: valentina.dompe@gssi.it

Abstract. Since the discovery of neutrino, many of its properties have been studied. The flavor oscillations recently revealed that, contrary to the Standard Model assumptions, neutrinos have a finite mass. Other neutrino characteristics are still to be investigated, such as its Dirac or Majorana nature: experiments are currently searching for a rare phenomenon, the neutrinoless double beta decay ($0\nu\beta\beta$), whose observation would solve this issue. The Cryogenic Underground Observatory for Rare Events (CUORE) at Gran Sasso National Laboratories focuses on the search of ^{130}Te $0\nu\beta\beta$ and it is the first ton-scale bolometric experiment ever realized for this purpose. The detector is hosted inside a custom cryostat and consists of 988 TeO_2 crystals, operated at a temperature of $\sim 10\,mK$. The CUORE goals in 5 years of data taking are an energy resolution of $5\,keV$ FWHM and a background level of 0.01 $counts/keV/kg/y$ in the ^{130}Te $0\nu\beta\beta$ region of interest. This corresponds to a sensitivity on the ^{130}Te $0\nu\beta\beta$ half life of $T_{1/2} = 9 \cdot 10^{25}\,y$ (90% C.L.) and an upper limit to the effective Majorana mass of $50 - 130\ meV$. The CUORE commissioning has been recently completed; the experiment has just concluded the pre-operation phase and data taking is currently ongoing.

1. Introduction

The experimental observation of neutrinos flavor oscillations is one of the most important results achieved in neutrino physics. The occurrence of this phenomenon implies that neutrinos are not massless; in this scenario, flavor and mass eigenstates are not coincident, since one can be expressed as linear combination of the others. Flavor oscillation can only happen if their masses are non-zero and different among each other [1]. This is the first experimental evidence of a physics beyond the Standard Model. There are many neutrino's characteristics which are still unknown, such as their absolute mass scale, the direct or inverse mass hierarchy, their Dirac or Majorana nature. Oscillations experiments cannot investigate on neutrino mass, since they are sensitive only to the mass difference, and not even on its nature, as the Majorana phases have no effects in neutrino oscillations [2]. An interesting approach for studying the neutrino nature is to search for the occurrence of the neutrinoless double beta decay ($0\nu\beta\beta$): it is a hypothesized decay mode of the double beta decay, a rare process in which the nucleus decays through a second order β such that two neutrons simultaneously go into protons. In the Standard Model, this is admitted with the emission of two neutrinos ($2\nu\beta\beta$), so that symmetries such as the lepton number are conserved regardless of the neutrino nature; this process has been already observed. However, if the neutrino were a Majorana particle, that is to say, if neutrino and antineutrino were coincident particles, the double beta decay would be allowed without neutrino emission.

The physical parameter involved in this process is the *effective Majorana mass*, defined as [3]:

$$\langle m_{\beta\beta}\rangle = \left|\sum_{j=1}^{3} U_{ej}^2 m_j\right| = \left|\sum_{j=1}^{3} |U_{ej}|^2 e^{i\phi_j} m_j\right| \tag{1}$$

where ϕ_j are the Majorana phases, and the $0\nu\beta\beta$ decay rate can be written as [3]:

$$[T_{1/2}^{0\nu}]^{-1} = G_{0\nu}(Q,Z)|M_{0\nu}|^2\langle m_{\beta\beta}\rangle^2 \tag{2}$$

where $G_{0\nu}$ is the phase space factor, $|M_{0\nu}|$ is the nuclear matrix element, $m_{\beta\beta}$ is the effective Majorana mass. This expression reveals that when lower limits on the lifetime are available, it is possible to set upper limits on $m_{\beta\beta}$.

The experimental approach for the detection of the double beta decay consists of measuring the energy of the emitted electrons. In case of a $2\nu\beta\beta$ event, there are overall four decay products, namely the two electrons and the two neutrinos: since the two neutrinos carry away a fraction of the energy, the spectrum of the sum of the two electrons energies turns out to be a continuum between 0 and the Q-value of the reaction. Conversely, in the $0\nu\beta\beta$ decay, there would be only two electrons in the final state, therefore they will carry the total amount of the decay energy (the kinetic energy due to the nuclear recoil is neglected). It immediately follows that the expected sum energy spectrum of the emitted electrons is a monoenergetic peak at the energy defined by the Q-value, and it has a non-zero width due to the finite detector resolution.

An essential parameter for the comparison of different experiments is the *sensitivity* S [3]. This quantity is defined as the half life corresponding to the minimum number of observable signal events above the background at a given statistical significance, or, in other words, to the maximum number of signal events hidden by the background fluctuations:

$$S^{0\nu}(n_\sigma) = T_{1/2}^{0\nu}(n_\sigma) = \frac{ln\,2}{n_\sigma}\,\epsilon\,N_a\,\frac{\eta}{A}\sqrt{\frac{M\,t}{b\,\Delta E}} \tag{3}$$

being n_σ the number of the standard deviations corresponding to a given confidence level (C.L.), ϵ the detection efficiency, N_a the Avogadro number, η the emitter isotopic abundance, A its mass number, M the total mass of the detector, t the experiment live time, b the background and ΔE the instrumental spectral width related to the energy resolution in the Region of Interest (ROI), that is an energy range around the Q-value of the $\beta\beta$ emitter.

Since this process is extremely rare, the awaited rate of events is very low, of the order of 1 event per $10^{25} - 10^{26}$ years. This is the main issue that makes its identification so difficult: in fact, experiments searching for rare events are strikingly sensitive to the background, and it is very difficult to discriminate it from signal events. There exist many different background sources that can hide a hypothetical $0\nu\beta\beta$ signal: the internal background sources coming from the detector itself, such as $2\nu\beta\beta$ spectrum or crystals radioactive contaminations, and the external ones, like muons, neutrons and γ from cosmic rays or rocks natural radioactivity. The background sources minimization therefore plays a crucial role in a detector searching for rare events.

2. The CUORE experiment

The Cryogenic Underground Observatory for Rare Events (CUORE) is currently the largest cryogenic bolometric detector searching for the $0\nu\beta\beta$ decay, in particular of the ^{130}Te isotope [4]. It is located at the Gran Sasso National Laboratories (LNGS) of INFN in Assergi (AQ), Italy, built under 1400 m of rock inside the Gran Sasso mountain. This allows to considerably reduce the rate of cosmic rays background. The muon flux is reduced by about six orders of magnitude, as it is $\sim 1\ \mu/(h \cdot m^2)$.

2.1. Bolometric technique & setup

A bolometer is a Low Temperature Detector (LTD) that exploits the temperature variation induced by a particle passing through it, in order to measure the amount of energy lost by the particle itself. The amplitude of the thermal signal ΔT can be expressed as a function of the deposited energy ΔE and the heat capacity C:

$$\Delta T = \frac{\Delta E}{C} \tag{4}$$

It is clear from this expression that the smaller is C, the higher is the signal amplitude; therefore, in order to have a good resolution, it is necessary to get a small value of C. It is well known that, being the absorber a dielectric and diamagnetic crystal, its thermal capacity follows the Debye law:

$$C(T) = \beta \frac{m}{M} \left(\frac{T}{T_D} \right)^3 \tag{5}$$

where $\beta = 1944\ J/(K \cdot mol)$, m is the crystal mass, M is the molecular weight and T_D is the Debye temperature. This expression tells us that the heat capacity C decreases as T^3 as the temperature also decreases; this means that a low temperature provides a low value of C, and therefore a higher signal amplitude. Thanks to this detector characteristic, it is possible to get a measurable temperature variation although the amount of energy deposition is very small: at room temperature, the thermal variation would be of the order of $10^{-18} - 10^{-15}\ K$, so it would be impossible to detect. The working temperature that allows bolometers to measure ΔT of the order of $\sim 10 - 100\,\mu K$ usually lies in the range $10 - 100\ mK$.

The CUORE experiment is based on the use of TeO_2 cryogenic bolometers, which represent both the source and detector of the ^{130}Te $0\nu\beta\beta$ decay; the isotope ^{130}Te is characterized by a natural abundance of 34.2%, which is a very important feature since it allows to realize an experiment by using only natural Te, avoiding to carry out expensive enrichment procedures that may also introduce radioactive contamination into the crystal. The searched process is described by the following reaction:

$$^{130}Te \rightarrow ^{130}Xe + 2e^- \tag{6}$$

that has a Q-value of $2527.515 \pm 0.013\,keV$ [5].

The CUORE setup consists of 988 TeO_2 crystals, which are arranged in 19 towers of 52 absorbers each [6] for a ^{130}Te mass of 206 kg out of 742 kg of total Te mass. A copper structure, working as heat bath, is mechanically and thermally coupled to the crystals through Teflon (PTFE) links. The temperature variation of the CUORE bolometers is converted into an electrical signal by Neutron Transmutation Doped (NTD) germanium semiconductor thermistors, which are glued on the crystals' surface. Bolometers are strikingly sensitive to thermal drifts, that can affect the detectors response and worsen their energy resolution. In order to avoid this phenomenon in CUORE, a $2.3 \times 2.4 \times 0.5\ mm^3$ silicon resistor glued on each crystal, the *heater*, is used to periodically inject a well known energy pulse into the crystals, in order to simulate physics events. Being the energy of the injected pulse constant, it allows to control the gain of every bolometer and correct the effects of thermal drifts during the first level of the offline data analysis through a stabilization procedure.

The CUORE towers have been assembled and installed in a clean room in the underground LNGS laboratories. The most delicate operations, such as the towers assembly, have been carried out into specific glove boxes kept under Nitrogen atmosphere to further reduce the contamination; a view of the CUORE towers storage and their completed installation are shown in Figs. 1 and 2, respectively.

IOP Conf. Series: Journal of Physics: Conf. Series **956** (2017) 012011 doi:10.1088/1742-6596/956/1/012011

Figure 1. The CUORE towers storage on Nitrogen overpressure, before their installation.

Figure 2. A front view of the CUORE towers, immediately after their installation.

2.2. The CUORE cryostat

The CUORE detector, operating at a temperature of $\sim$ 10 mK, is housed into a dedicated custom-built cryostat, that is one of the largest ever realized [7]. In order to avoid material damages as consequences of the sudden thermal jump, the base temperature is reached through different stages of cooling. The CUORE cryostat is therefore composed of six copper shieldings that are thermalized to decreasing temperatures: $300\,K$, $40\,K$, $4\,K$, $600\,mK$, $50\,mK$, $10mK$ (Mixing Chamber). The intermediate vessels work as radiation shields, so that thermal radiations from the warmer stages cannot affect the detectors. In order to further strengthen this thermal insulation, the 40 K and the 4 K vessels are both vacuum tight at $< 10^{-8}\,mbar$, moreover they are wrapped in the so-called *superinsulation*, which is a multi-layer mylar insulation.

Vacuum and superinsulation minimize the heat radiation reaching the detector, but do not protect it from the external radioactivity: lead shields housed inside the cryostat aim at this task. A 30 cm thick top shield of 2745 kg lead is placed above the Tower Support Plate (TSP) and is thermalized to 50 mK, while a 5562 kg lateral shield of ancient Roman lead envelopes the cryostat between the 4 K and the 600 mK vessel and is thermalized to the 4 K vessel because of stronger cooling power. In 1988, 120 lead ingots were recovered from a Roman shipwreck that sank around 80-50 B.C. off the coast of Sardinia. The relic was preserved at the National Archeological Museum in Cagliari, and in 2010 30 ingots were tranfered at LNGS thanks to an agreement between the INFN and the Italian cultural heritage Ministry [8]. This ancient lead has spent 2000 years under the sea, so nowadays it has almost completely lost its natural radioactivity: this makes it the perfect material for a CUORE shielding.

Two different techniques are used for the CUORE cryostat cool down [4]. The Outer Vacuum Chamber (OVC), namely the 40 K and 4 K vessels, is cooled by 5 Pulse Tubes (PTs) coolers: with respect to conventional cryogenic diluition refrigerators, the PTs have no need of cryogenic liquids, therefore they do not require periodical refilling interruptions that affect the experiment live time.

IOP Conf. Series: Journal of Physics: Conf. Series **956** (2017) 012011 doi:10.1088/1742-6596/956/1/012011

From 4 K, the base temperature is reached through a continuous-cycle Diluition Refrigerator (DR), a cooling system that exploits the properties of the ^{3}He-^{4}He mixture: at few degrees above the absolute zero, ^{3}He and ^{4}He cannot arbitrarily mix, they instead spontaneously separate into two phases: the *concentrated phase*, mainly containing ^{3}He, and the *diluite phase*, mostly containing ^{4}He. The DR uses a pump external to the cryostat that pulls ^{3}He from the concentrated phase to the diluite phase, and this produces an endothermic process that, absorbing the energy, allows the cryostat to achieve the base temperature. The ^{3}He is finally returned to the diluite phase, closing the cycle. This cool down system requires about 2 days for a 15 tons total mass such as that of the CUORE cryostat.

A schematic image of the CUORE cryostat is shown in Fig. 3.

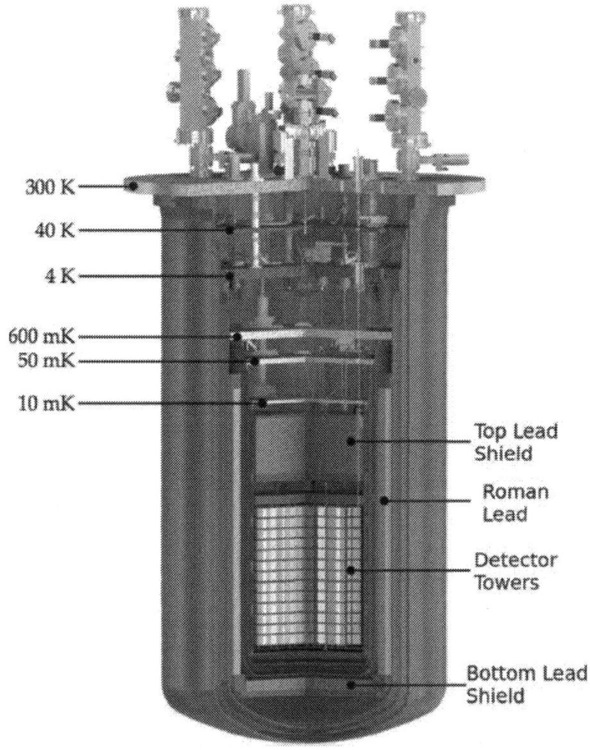

Figure 3. A scheme of the internal structure of the CUORE cryostat. The copper shields and their temperatures are shown, the towers being hosted into the inner one; the lead shields for the radiation protection are also represented.

3. CUORE goals

Thanks to the CUORE experimental characteristics such as its large exposure, the appropriate TeO_2 crystals features at cryogenic temperatures and the specific procedures adopted for cleaning and assembling the detector components, in order to minimize the internal background, a competitive result will be likely set for the $0\nu\beta\beta$ decay of ^{130}Te. CUORE aims at obtaining a FWHM energy resolution of $\sim 5\ keV$ at the $0\nu\beta\beta$ energy region of interest of ^{130}Te. The dedicated cleaning operations and the new cryostat, built starting from radiopure selected materials, will allow to reduce the background level to $10^{-2}\ counts/keV/kg/y$ in the same energy region; the ^{130}Te half life sensitivity for CUORE is expected to be $9 \cdot 10^{25}\ y$ (90% C.L.), which corresponds to an upper limit range for the Majorana neutrino mass of $50 - 130\ meV$. Fig. 4 shows the sensitivity that CUORE would be able to reach for ^{130}Te.

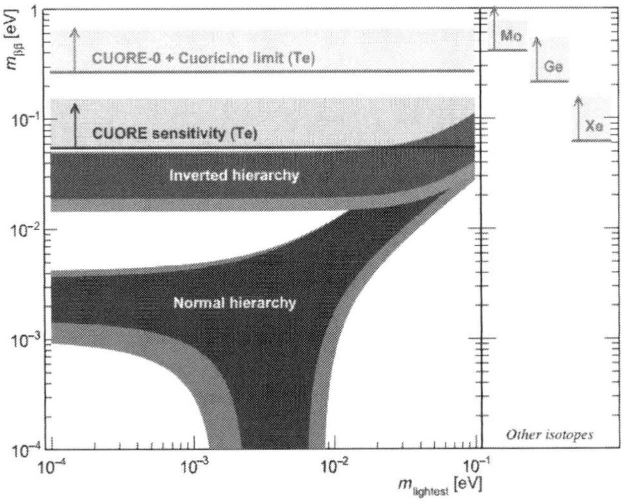

Figure 4. Effective Majorana mass $m_{\beta\beta}$ vs lightest neutrino mass [9]. On the right side of the plot, limits set by other experiments on other isotopes are also shown.

4. Conclusions

The CUORE commissioning phase started in spring 2016, with the cryostat operations; just few months later, in July - August, the towers installation was completed and the cryostat copper shields were closed in November. The cooldown phases started in December 2016 and were successfully completed with the reach of the base temperature at the beginning of January 2017; just a few days later, the first CUORE pulses have been observed, confirming the proper functioning of the detectors. Pre-operation procedures, such as working parameters optimization followed, and the calibration phase took place in April. The commissioning phase has then been completed within the end of April; CUORE data taking has started and will continue for 5 years.

References

[1] Patrignani C *et al.* Particle Data Group 2016 *Chin. Phys. C* **40** 100001
[2] Giunti C 2010 *Phys. Lett. B* **686** 1 41-43
[3] Avignone F T, Elliott S R and Engel J 2008 *Rev. Mod. Phys.* **80** 481
[4] Ligi C *et al.* 2016 *Low Temp. Phys* **184** 3 590-596
[5] D'Abbabbo A *et al.* 2016 *Preprint* arXiv:1612.04276
[6] Artusa D R *et al.* 2015 *Adv. High Energy Phys.* **2015**
[7] Singh V *et al.* 2016 *J. Phys.: Conf. Ser.* **718** 062054
[8] Nosengo N Apr 2010 *Nature News*
[9] Strumia A and Vissani F 2005 *Nucl. Phys. B* **726** 294-316

IOP Conf. Series: Journal of Physics: Conf. Series **956** (2017) 012012 doi:10.1088/1742-6596/956/1/012012

3D detectors with high space and time resolution

A Loi[1,2]

[1] Università di Cagliari, Dipartimento di Fisica, Complesso Universitario di Monserrato - S.P. per Sestu Km 0.700, 09042 Monserrato (Cagliari), Italy
[2] INFN-Sezione di Cagliari, Complesso Universitario di Monserrato - S.P. per Sestu Km 0.700, 09042 Monserrato (Cagliari), Italy

E-mail: angelo.loi@ca.infn.it

Abstract. For future high luminosity LHC experiments it will be important to develop new detector systems with increased space and time resolution and also better radiation hardness in order to operate in high luminosity environment. A possible technology which could give such performances is 3D silicon detectors. This work explores the possibility of a pixel geometry by designing and simulating different solutions, using Sentaurus Tecnology Computer Aided Design (TCAD) as design and simulation tool, and analysing their performances. A key factor during the selection was the generated electric field and the carrier velocity inside the active area of the pixel.

1. Introduction

In future high luminosity LHC (HL-LHC) experiments planned over the next 10 years, the number of collisions per unit of time is expected to increase by more than a factor of 2 [1]. This will make possible to study rare phenomena and extend the knowledge of fundamental interactions. To detect particle collisions, their products and also the interaction processes in these extreme conditions, detectors with very high space and time resolution will be needed (less than 100 μm in space and 100 ps in time). To date, there are no particle detectors with such performances. A new approach to improve especially the time resolution is to use three-dimensional (3D type) detectors. 3D Detectors are solid state devices introduced for the first time

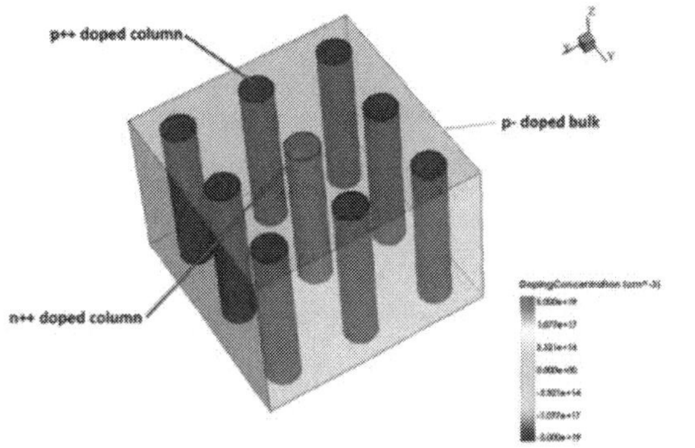

Figure 1. Schematic view of a typical 3D-detector pixel. In this model the electrodes have a cylindrical design and cross the wafer through all the thickness.

by Sherwood Parker during the nineties [2] and have been used increasingly in particle physics experiments in the last decade, like at ATLAS IBL [3]. Unlike classic planar detectors, the main feature of these devices is to have their electrodes oriented perpendicularly to the wafer surface (Fig. 1) and penetrating deep inside the bulk (up to hundreds of μm). The main structural advantage of this approach is that the distance between the electrodes is decoupled from the wafer thickness, which gives the detector a lower bias voltage, rapid charge collection and so a faster time response (less than 500 ps) and a higher current signal amplitude [4]. High intrinsic radiation tolerance, due to the shorter inter-electrode distance and the lower bias voltage, is another main feature of those devices, demonstrated with an intense proton irradiation [5].

The realization of this kind of particle detectors requires non-standard realization processes that allows to excavate with accuracy for several hundreds of μm without damaging the wafer. Main techniques used actually by the foundries to build such structure are, for example, etching techniques like DRIE (Deep Reactive Ion Etching)[6].

The possibility to use the shorter charge collection time of 3D detectors for timing application has not yet been thoroughly investigated. First studies about timing performances and possible associate fast front-end electronics are analysed by Sherwood Parker et al. in 2011 [4].

The aim of this work is to study the possibility of a particle detector with high time and space resolution and also high radiation hardness in order to operate in the extreme conditions in future HL-LHC experiments, starting with a first selection of possible pixel geometries.

2. Geometry Design and Simulation

At first more than 20 different geometry solutions have been explored and two of them have been selected for a more accurate analysis. The geometries are divided up into 2 main categories: square pixel geometries with a dimension of 50 μm times 50 μm and hexagonal pixel geometries with 50 μm in diameter. With these dimensions the charges reaches in less time the electrodes, reducing charge collection time and thus increasing the time resolution. All solutions share the same constant doping profiles, with concentrations of $5{*}10^{19}/\mathrm{cm}^3$ for the p++ (Boron) and n++ (Arsenic) doped electrodes and $7{*}10^{11}/\mathrm{cm}^3$ for the p−− doped bulk (Fig. 2). Design and

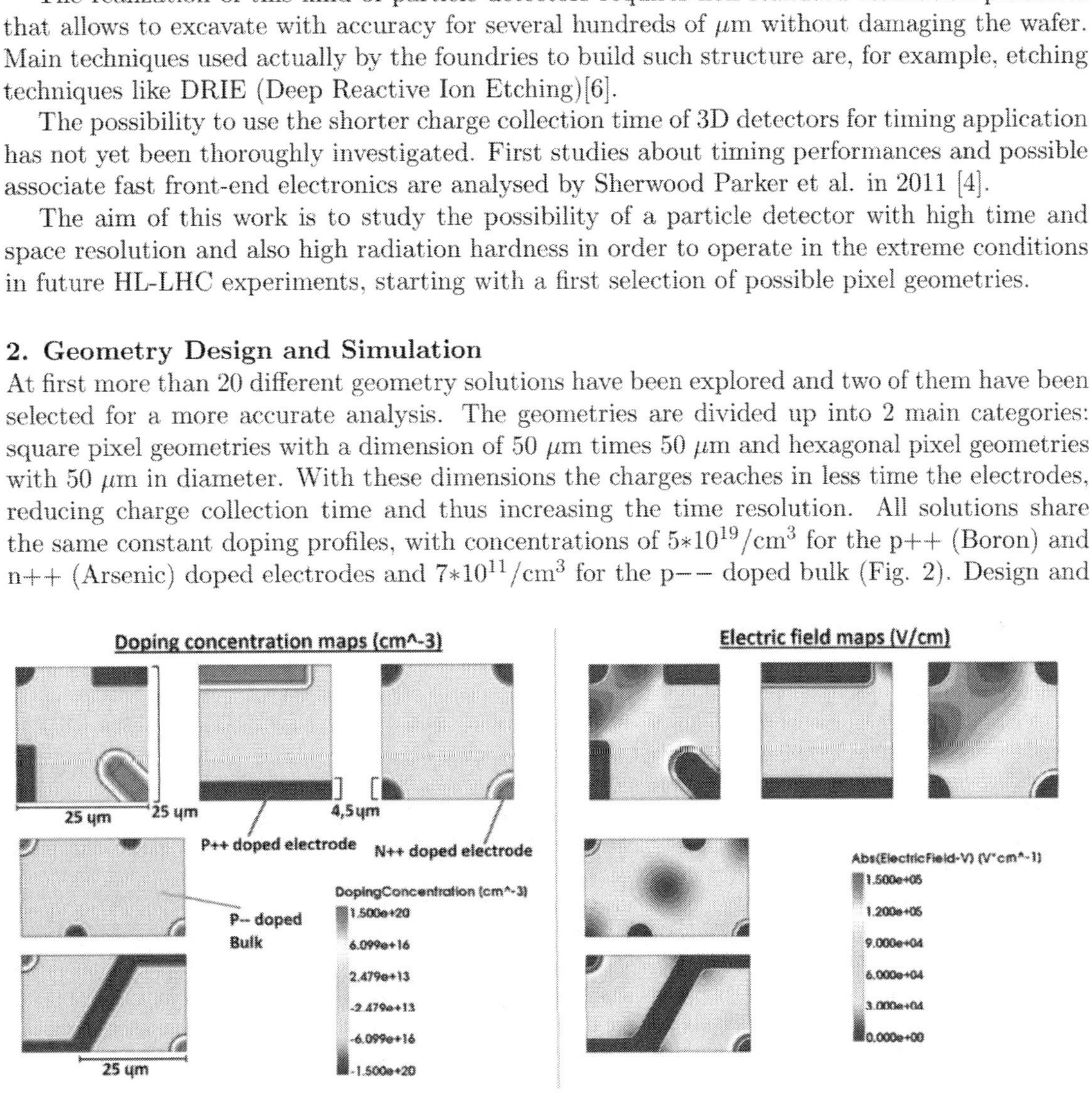

Figure 2. Doping concentration and electric field maps of 5 of the 20 explored geometry solutions. Hexagonal and square pixels are not in same scale.

simulation were performed using Synopsys Sentaurus TCAD [7] to predict mainly the electric field and the carrier velocity inside the active area of the device. Those informations are found by solving the continuity and Poisson's equations, which relates charge density and potential.

Also the simulation takes into account all charges inside the device and possible presence of traps, level defects and carrier recombinations [8]. Simulation was set in quasi stationary mode with a P-electrode bias voltage ramp starting from 0 V and decreasing to -200 V, while the N-electrode potential was set constant at 0 V. During the simulation, electric field and carrier velocity maps were saved at voltages of -50 V, -100 V, -150 V and -200 V. The second bias voltage was chosen as reference voltage to compare the performances of all geometries. During the first selection step only 2D models of the geometries were designed in order to understand the electric field behaviour between the electrodes and the velocity for electrons and holes inside the devices. This approach, including 2D models which represents only 1/4 of the entire pixel, reduces computational resources and simulation time thus increasing the number of explored solutions. The analysis of the electric field was focused mainly by inspecting its amplitude in relation to the position inside the active region, which needs to be around 10-100 kV/cm, conditions to which velocity saturation occurs. In this way it is guarantee that the output signal of the detector has a constant shape independent of the position. This is possible thanks to the charge carrier velocity which is at its maximum value and stays constant. For pure silicon $1.07 * 10^7 cm/s$ for electrons and $8.37 * 10^6 cm/s$ for holes at 300 C[8].

Particular attention was paid to control the behaviour of the electric field in the critical regions, for example the space between electrodes at same potential. In these areas the electric field decreases too much and signal generation could potentially become too inefficient. In this case it is important that the electric field amplitude stays high enough to guarantee velocity saturation. If it is not possible to ensure velocity saturation inside those regions, charge collection becomes slower, making the device practically useless for the requested high performances. From

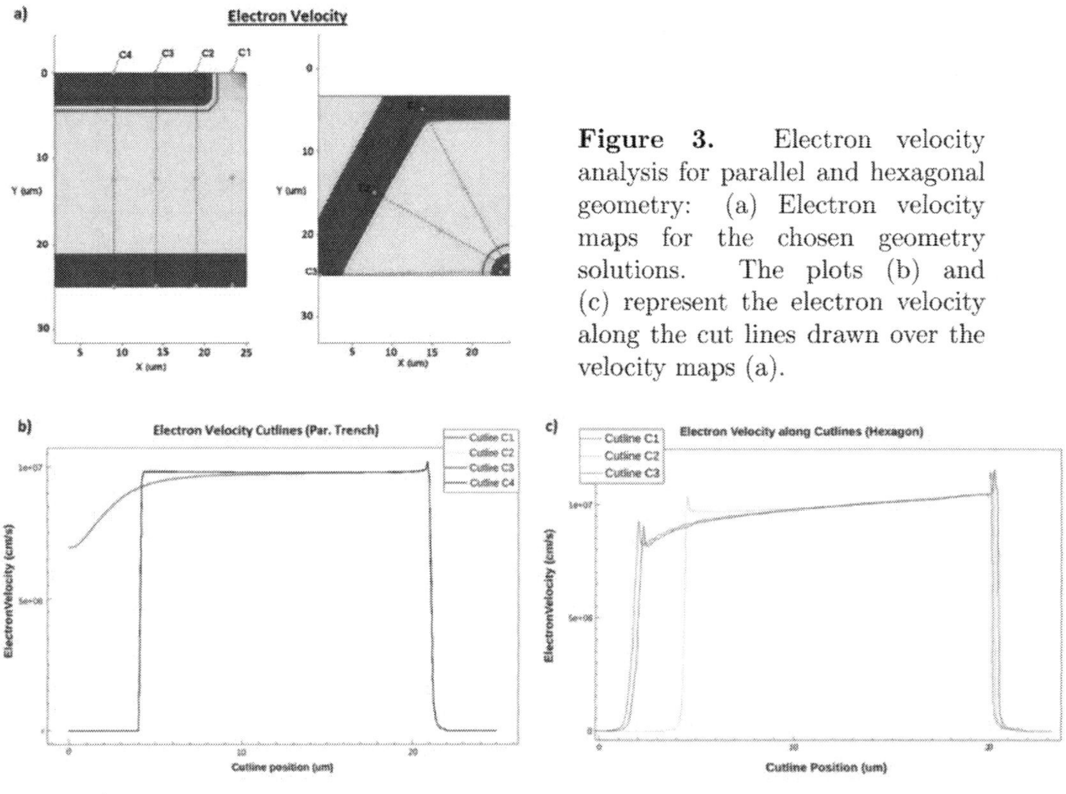

Figure 3. Electron velocity analysis for parallel and hexagonal geometry: (a) Electron velocity maps for the chosen geometry solutions. The plots (b) and (c) represent the electron velocity along the cut lines drawn over the velocity maps (a).

this first analysis only two solutions were chosen for the next step: one square pixel design with parallel trench geometry and one hexagonal solution with closed trench geometry. The parallel trench pixel solution was chosen because it has an uniform electric field over all the active area. In the region between two n++ doped electrodes carrier velocity is below saturation level but can be neglected because this region is close the electrodes, ensuring a fast charge collection. Instead of the parallel trench, the hexagonal solution presents a less uniform field but the velocity saturation covers almost the entire active area, except the angular regions, where charge velocity goes under saturation level (Fig. 3). It is possible to restore the saturation in those regions by increasing depletion voltage from 100 V to 150 V.

During the second step, three dimensional models of the chosen pixels were designed and simulated. The models include more details of the real devices like p-spray surface doping, details of the electrode structure, silicon oxide layers and also metallic contacts. In this way it is possible to achieve a deeper understanding of the electric field behaviour inside the device. The results obtained from the simulation of the two models have shown that the electric field inside the parallel square geometry changes heavily along the bulk surface. Especially a high electric field was observed along the upper surface between the n++ doped electrode and the p-spray layer (Fig. 4). At the bottom of the pixel, the electric field decreases too much and charge velocity goes under its saturation value. A similar behaviour has also been observed for the hexagonal geometry, suggesting to find solutions in order to minimize this problem.

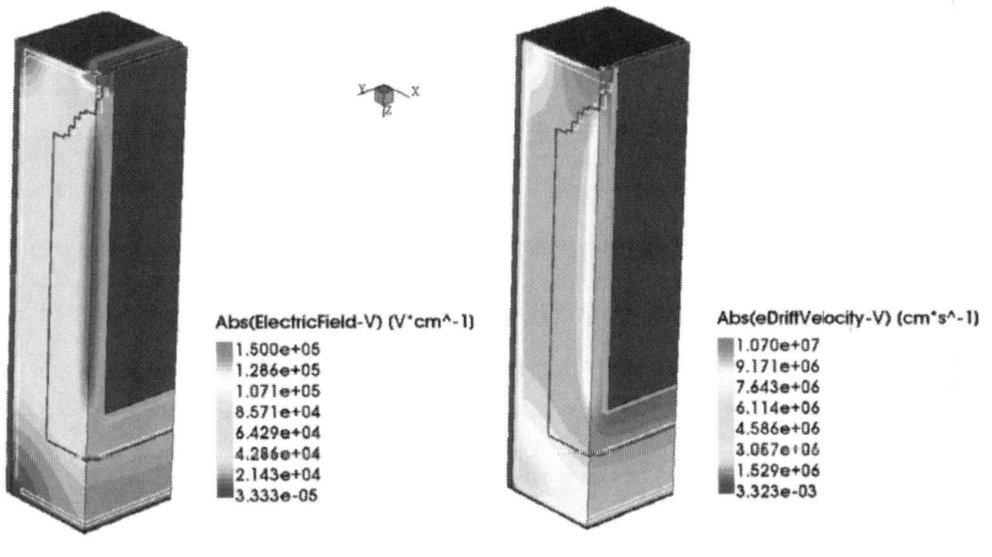

Figure 4. Electric field (left) and electron velocity (right) maps for the 3D model of the parallel trench pixel design. Velocity saturation is not maintained over all the active area, only between the main electrodes..

3. Next steps

The possibilities to simulate, using TCAD, the operation of the detector when crossed by a high energy particle are limited. This is due to the fact that the software is more oriented in the development and simulation of semi-conducting devices for mass production such as MOSFETs or LEDs, instead of particle detectors for high energy particle physics. The only way to simulate in first approximation the passage of a high energy particle through the detector and observe the generated current signal is to use the Sentaurus device Heavy Ion model (HI) [8], which is an implemented physical process developed for the main purpose of studying the effects of charge deposition from a high energy particle impact, like Single Event Upsets (SEU). It is

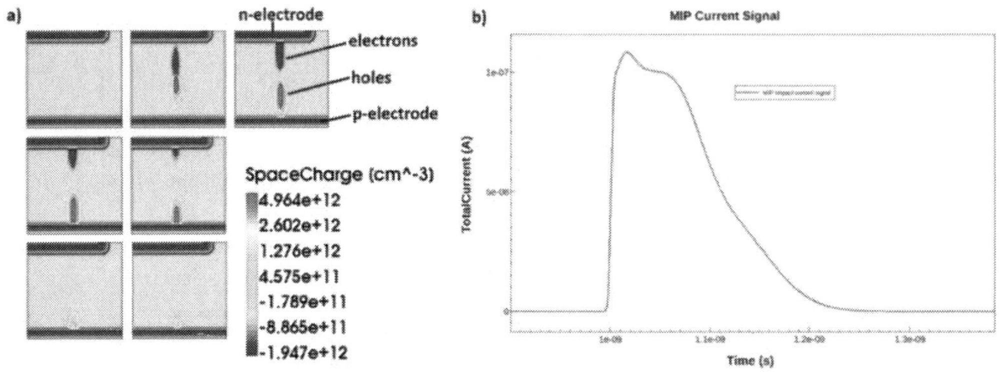

Figure 5. (a) Sequence showing the passage of electrons and holes generated by a passing MIP along the middle of the chosen pixel and respective current signal (b).

possible to tune an HI for many different cases, changing its specific parameters, which are the coordinates of the starting point, the direction and the length of the path and also the Linear Energy Transfer (LET) of the heavy ion [8]. In this case, the LET was set at 80 pair/μm in order to emulate the passage of Minimum Ionizing Particle (MIP), which can be for example a high energy proton. Using this kind of radiation model, only primary emission charge is considered, which is useful only to have a first charge collection time estimation.

Second emission particle, like delta rays, which are very important in understanding the timing error introduced by the detector, as mentioned by Sherwood Parker [4], are not considered in the HI model.

HI interaction was simulated in transient mode, saving every 30 ps a space charge map (Fig. 5a). The results obtained with this approximated approach showed current signals with 200-400 ps duration and rise time of 15-30 ps, in the case of parallel trench geometry [Fig. 5b].

Considering that the TCAD simulation is less accurate for describing a particle detector and its operation, a specific Monte Carlo simulation is under development, which will use the exported electric field maps from TCAD and will implement further interaction processes, like delta ray emission, in order to have a more accurate view of the time performances of the device. An other approach, based on Geant4 is currently under evaluation.

4. Conclusions

A first selection of detector pixel geometries has been completed and 2 main designs have been chosen (a square and a hexagonal). These geometric solutions have the best characteristics in terms of electric field amplitude and coverage, allowing to work in velocity saturation regime (with some limitations), which is the recommended operating configuration in order to keep an output signal with a specific shape.

First particle strike simulations show that the output signals were of the order of hundreds of ps with rise times around 20 ps.

Considering the lack of information about the performances of the detector in working conditions, it is important to investigate this issue by using other approaches, such as implementing a Monte Carlo simulation or using specific simulation tools like Geant4.

Acknowledges

I would like to thank Prof. Gianfranco Dalla Betta and Dr. Roberto Mendicino of the University of Trento for introducing me to 3D-detector design and simulation and for their initial support and guidance in this complex discipline.

References

[1] Schmidt B 2015 *The High-Luminosity upgrade of the LHC*, J.Phys.Conf.Ser. **706** no.2, 022002
[2] Parker S I, Kenney C J and Segal J 1997 *3D - A proposed new architecture for solid-state radiation detectors*, Nucl. Instrum. Meth A **395** 328-343
[3] ATLAS Collaboration 2010, *ATLAS Insertable B-Layer Technical Design Report*, ATLAS TDR 19, CERN/LHCC **2010-013**
[4] Parker S I *et al.* 2011 *Increased Speed: 3D Silicon Sensors; Fast Current Amplifiers* IEEE Trans. Nucl. Sci NS-**58**(2) 404
[5] Parker S I and Kenney C J 2001 *Performance of 3-D Architecture Silicon Sensors After Intence Proton Irradiation* IEEE Trans. Nucl. Sci NS-**48**(5) 1629
[6] Wu B *et al.* 2010 *High aspect ratio silicon etch: A review* J. Applied Physics **108** 051101
[7] Synopsys Inc. 2015 *Synopsys Sentaurus TCAD* Version 2015.06, https://www.synopsys.com/silicon/tcad.html
[8] Synopsys Inc. 2015 *Synopsys Sentaurus Device User Manual* Version 2015.06

MONDO: A tracker for the characterization of secondary fast and ultrafast neutrons emitted in particle therapy

R Mirabelli[1,2], G Battistoni[3], V Giacometti[2,4], V Patera[2,4,5], D Pinci[2], A Sarti[4,5,6], A Sciubba[2,4,5], G Traini[1,2], M Marafini[2,4]

[1] Dipartimento di Fisica, Sapienza Università di Roma, Roma, Italy
[2] INFN Sezione di Roma, Roma, Italy
[3] INFN Sezione di Milano, Milano, Italy
[4] Centro Fermi, Museo Storico della Fisica e Centro Studi e Ricerche "E. Fermi", Roma, Italy
[5] Dipartimento di Scienze di base e Applicate per l'Ingegneria, Sapienza Università di Roma, Roma, Italy
[6] INFN Sezione Laboratori Nazionali di Frascati, Frascati, Italy

E-mail: riccardo.mirabelli@roma1.infn.it

Abstract. In Particle Therapy (PT) accelerated charged particles and light ions are used for treating tumors. One of the main limitation to the precision of PT is the emission of secondary particles due to the beam interaction with the patient: secondary emitted neutrons can release a significant dose far from the tumor. Therefore, a precise characterization of their flux, production energy and angle distribution is eagerly needed in order to improve the Treatment Planning Systems (TPS) codes. The principal aim of the MONDO (MOnitor for Neutron Dose in hadrOntherapy) project is the development of a tracking device optimized for the detection of fast and ultra-fast secondary neutrons emitted in PT. The detector consists of a matrix of scintillating square fibres coupled with a CMOS-based readout.

Here, we present the characterization of the detector tracker prototype and CMOS-based digital SPAD (Single Photon Avalanche Diode) array sensor tested with protons at the Beam Test Facility (Frascati, Italy) and at the Proton Therapy Centre (Trento, Italy), respectively.

1. Introduction

Nowadays Particle Therapy (PT) is more and more diffused as proven by the number of treatment centers with light ions (mainly protons and carbon ions) that are continuously increasing. The main advantage of PT, compared to conventional radiation therapy, is its power in concentrating the dose release in the tumor region preserving the surrounding tissues. The charged particles, protons and light ions, release the maximum dose at the end of their path in the well known Bragg Peak: the position of that peak, as well as its width, depends on beam energy and on type of particles used. The abundant components of secondary particles (neutral and charged particles) produced by the nuclear fragmentation of the beam (and of the target) and by the de-excitation of the isotopes exited by the beam interaction in the matter, has been previously studied [1-3]: charged particles, prompt photons and photons from $\beta+$ annihilation. In general, the additional dose due to charged secondary products and photons is negligible compared to the beam one. However, the secondary particles have been carefully characterised

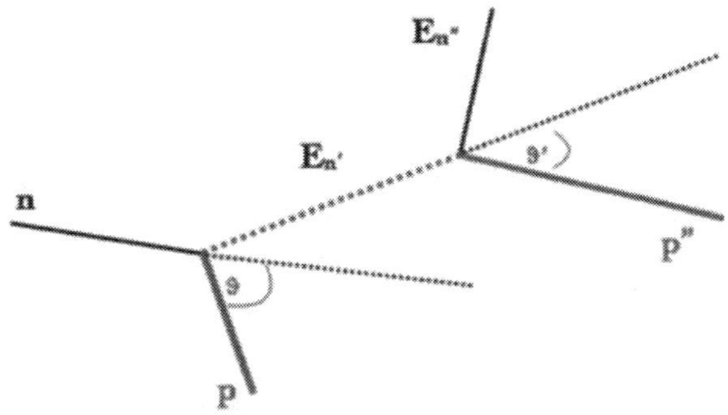

Figure 1. Schematic representation of a neutron double elastic scattering. Measuring protons tracks, kinetic energies and angles, it is possible to calculate the kinematics and establish direction and energy of the neutrons.

because of their usefulness in beam range monitoring and partially contribute to the effective dose delivered in treatments. For the neutron component instead the characterization is not complete: fluxes and energy spectra in function of the emission angle have been measured for thin target [4] and no efficent tracking detector for neutrons are nowadays available in the energy range of PT interest. Available measurements are based on the combination of different experimental setups, e.g. with an extended range Bonner sphere [5] or on the use of detectors with long and complex calibrations. Secondary neutron production is one of the main cause of Secondary Malignant Neoplasms (SMN), which develop significantly later (many years) after the treatment [6]. Therefore, it is crucial to characterise the secondary neutron production in the patient and the limit of the application of PT in the treatment of paediatric tumors is in this context evident.
The aim of the MONDO project (MOnitor for Neutron Dose in hadrOntherapy) is to characterise the secondary neutrons production in PT in order to improve the Monte Carlo and the analytical models at the base of the Treatment Planning System (TPS) [7]. A tracker dedicated to the measurements of neutrons in the energy range of 20 - 400 MeV will be built. Specifically, Sec. 2 presents the detector principle of operation, while some results with electrons and protons beam are reported in Sec. 3 testing a detector prototype.

2. The MONDO Project

In the MONDO tracker the complete reconstruction of the neutrons quadri-momentum is performed through the detection of recoiled protons emitted by neutrons Double Elastic Scattering (DES) with the interacting material. When two consequent recoiled protons emitted in DES are completely contained in the active area of the detector, the energy and the direction of the incoming neutrons are calculated using the following equation:

$$E_n = \frac{E_p}{cos^2\theta} \tag{1}$$

where E_i $(i = n, p)$ is the neutron/proton kinetic energy and θ is the angle between the recoil proton and the incoming neutron. Fig. 1 shows an example of a DES event.

The final design of the detector includes a $10 \times 10 \times 20$ cm^3 matrix of scintillating plastic fibers (square cross section, 250 μm side) arranged in orthogonal layers. The active volume acts both as a target for neutrons and as a detector for secondary emitted protons. The light produced by the proton interaction with the fibers is collected with a dedicated system developed

Figure 2. PENELOPE: a MONDO tracker prototype.

in collaboration with Fondazione Bruno Kessler (FBK).
The readout consists of a SPAD-array built with the same technology presented in [8], but with some characteristics optimized for our goal. In particular, the sensor trigger logic is based on a double threshold logic with a smaller pixel size (250×125 μm).
In order to characterise the fibers and to finalize the construction of the MONDO detector, a small prototype (called PENELOPE) was built (total dimension: $4 \times 4 \times 4.8$ cm^3). The device is shown in Fig. 2.

3. Experimental results

PENELOPE was tested both with electrons and protons; its efficiency and light production were characterised. The prototype was readout with three different devices: (1) a commercial Hamamatsu H10580 PhotomultiplierTube (PMT), (2) a Multianode H8500C and (3) a spad-net sensor (described in the previous Section).

3.1. Test with an electron beam

In May 2016 PENELOPE was tested at the Beam Test Facility (BTF) at the Frascati INFN National Laboratory. The goal of the test beam was the evaluation of the charged produced by a 450 MeV electron crossing a layer of fiber. In fact, at that energy electrons act like a mip (minimum ionizing particles) thus allowing to study the worst case scenario in energy release and light production. The Hamamatsu PMT response is presented here. Fig. 3 shows a schematic representation of the experimental setup and the average PMT output signal. The integral of the signal corresponds to the total charged produced and 250 ± 50 photoelectrons were measured. Considering that the PMT covers an equivalent area of 30 ± 2 layers, the calculated charged produced by a single layer is 7.2 ± 1.4.

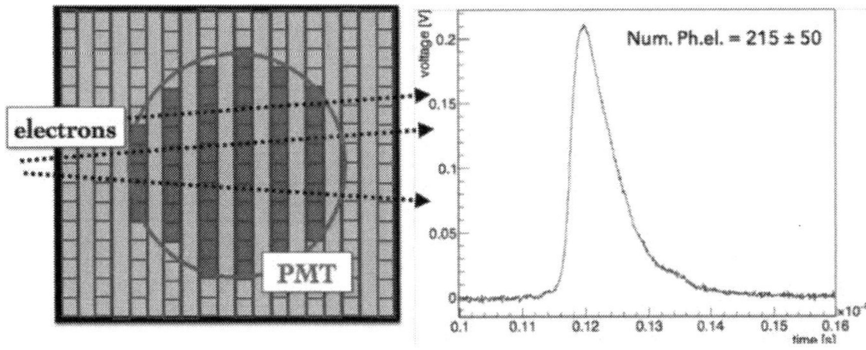

Figure 3. (Left) Experimental setup at the BTF: the area covered by the PMT is smaller than the total area covered by PENELOPE. (Right) Average waveform signal produced by the PMT. [9]

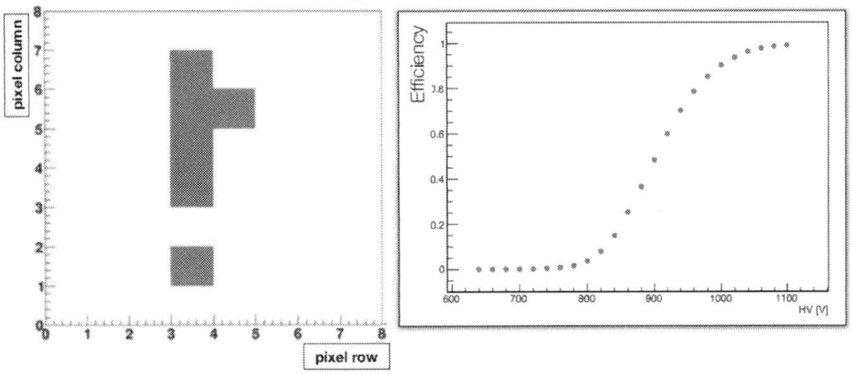

Figure 4. (Left) Example of an event with a track, in which 6 pixels in a row are activated. (Right) Efficiency in function of the multianode HV.

3.2. Test with a proton beam
In May 2017 PENELOPE was tested at the Proton Therapy Center (PTC) at the Trento Hospital, using protons in the energy range of 70 - 230 MeV, which corresponds to the energy spectra of the recoiled protons produced in DES neutron events. The multianode and spad-net sensor responses are presented here.

3.2.1. Multianode readout: efficiency A study of the efficiency in function of the multianode High Voltage (HV) supply was performed. The multianode H8500C has 64 pixel (8×8, 6 mm side) individually readout covering a total area of $\sim 5 \times 5$ cm^2. The efficiency is calculated as the number of events with at least 6 pixels in the same row showing a signal (i.e. a proton track) over the total number of events (Fig. 4).

The efficiency increases with HV. So we decided to operate our multianode with the maximum HV suggested by the multianode data-sheet.

3.2.2. Spad-net sensor: tracks reconstruction An analysis of the proton tracks extracted using the spad-net sensor was also performed: the single event maps, obtained counting the number

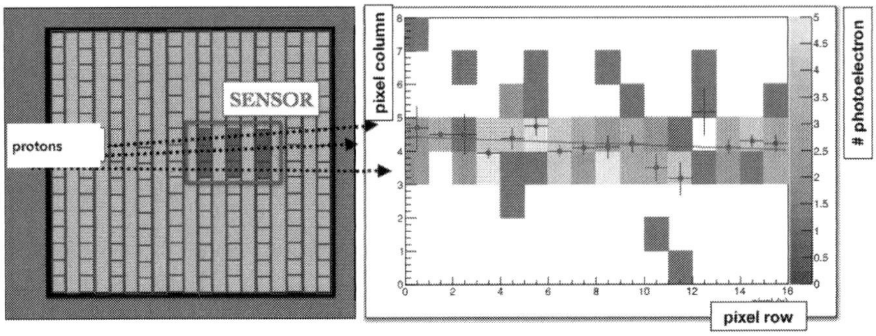

Figure 5. (Left) Expermental setup at the PTC: the area covered by the spad-net sensor is very small compared to the total area covered by PENELOPE. (Right) Example of a proton track reconstruction.

of photo-electrons detected from each pixel, were fitted with a linear function to reconstruct the proton direction (Fig. 5). In order to reduce the background noise, pixels with less than 2 activated spads were not included in the fit.

4. Conclusion

The MONDO project is dedicated to the characterization of secondary neutrons produced during particle therapy treatments. The system is based on the detection of double elastic scattering events (neutron on scintillating fibres) and the readout requirements are quite stringent.

The data colleced at BTF and CPT revealed that a fully digital CMOS approach is suitable for the MONDO tracker. The prototype shows good characteristics in efficiency, track reconstruction, light production and collection. A new readout sensor (SBAM: SPAD-Based Acquisition readout for MONDO experiment) is under development: the new design will match the tracker geometry and implement a smart discrimination mechanism specifically tailored to the proton tracks.

Acknowledgments

We would like to thank Marco Magi (SBAI Department) for his valuable work in the design and construction of the mechanical structure of the detector. The authors also thank the BTF and TIFPA staff for beam-time and helpful support during the data taking.

The MONDO project is supported by INFN Gruppo V with a Young Researchers Grant (2015-2016) and by SIR (Scientific Independence of young Researchers) MIUR Founding (2015-2018).

References

[1] Muraro S *et al.* 2016 *Review Article Front. Oncol.* **6**
[2] Mattei I *et al.* 2016 *Phys. Med. Biol.* **62** 1438
[3] Agodi C *et al.* 2012 *NIM B* **23** 1-8
[4] Gunzert-Marx K *et al.* 2008 *New Journal of Physics* **10**
[5] Mares C *et al.* 2016 *Phys. Med. Biol.* **61** 4127-4140
[6] Newhauser W and Durante M 2009 *Phys. Med. Biol.* **54** 2277-2291
[7] Marafini M *et al.* 2017 *Phys. Med. Biol.* **62** 3299-3312
[8] Braga L *et al.* 2014 *IEEE Journal of Solid-State Circuits* **49** 301
[9] Mirabelli R *et al.* 2017 *IEEE Transactions on Nuclear Science* Submitted

IOP Conf. Series: Journal of Physics: Conf. Series **956** (2017) 012014 doi:10.1088/1742-6596/956/1/012014

Impact of neutrino background prediction for next generation dark matter xenon detector

M Cadeddu[1,2] **and E Picciau**[1]

[1] Dipartimento di Fisica, Università degli Studi di Cagliari, S.P. per Sestu, 09042, Italy
[2] INFN-Sezione di Cagliari, Italy

E-mail: emmanuelepicciau1994@gmail.com

Abstract. Next generation direct dark matter detectors will have the sensitivity to detect neutrinos from several sources, among which atmospheric and diffuse supernova neutrinos, through the Standard Model reaction of Coherent Elastic Neutrino Scattering on nucleus. This reaction represents an irreducible background that can be expressed as a limit in the Weakly Interacting Massive Particles parameters plane. This limit is known as the "neutrino floor" and it has been obtained by other authors considering standard hypotheses for the neutrino-nucleus form factor and for the coherence of the scattering process. Since the coherent scattering has never been observed experimentally, it is licit to relax some hypotheses in the differential cross section and to evaluate the effect of such modifications on the neutrino floor prediction. In this contribution, we show a more accurate neutrino-nucleus form factor and we discuss the coherence hypothesis of the process in two extreme cases, namely the total coherence and the total decoherence regime. We derive the neutrino background event rate under these new assumptions, considering xenon as a target. The differences between the number of neutrino events and the implication for the next generation dark matter detectors, such as XENON1T/XENONnT, LZ and DARWIN, are discussed.

1. Introduction

The presence of Dark Matter (DM) plays a fundamental role at the cosmological and astrophysical scale. So called Weakly Interacting Massive Particles (WIMPs) are considered as the most promising DM candidates. Direct Detection (DD) experiments aim to identify WIMPs by measuring, with different technologies and target materials, the energy deposited by WIMPs after a nuclear recoil. One of the most used technique exploits the effect of scintillation and ionization of liquid nobles (such as xenon and argon) in Time Projection Chambers (TPCs) after a WIMP coherent interaction with nuclei. In this way, one can measure (constrain) two of the unknown WIMP parameters, namely the WIMP-nucleon cross section and the WIMP mass in the case of a positive (null) result. Nowadays, the best upper limit on the WIMP-nucleon cross section has been set by the LUX experiment [1], a Liquid Xenon (LXe) TPC and it is of the order of 10^{-46} cm^2 for a WIMP mass of about 30 GeV/c^2. In principle, increasing the exposure (the product of the active mass and the live time of the experiment) it is possible to explore lower and lower cross sections. However, it has been pointed out that neutrinos from several sources could scatter off coherently with the whole nucleus producing the same signal of a WIMP [2]. The Coherent Elastic Neutrino Scattering (CENS) process is relatively well described in the Standard Model (SM) framework but it has never been observed experimentally because

IOP Conf. Series: Journal of Physics: Conf. Series **956** (2017) 012014 doi:10.1088/1742-6596/956/1/012014

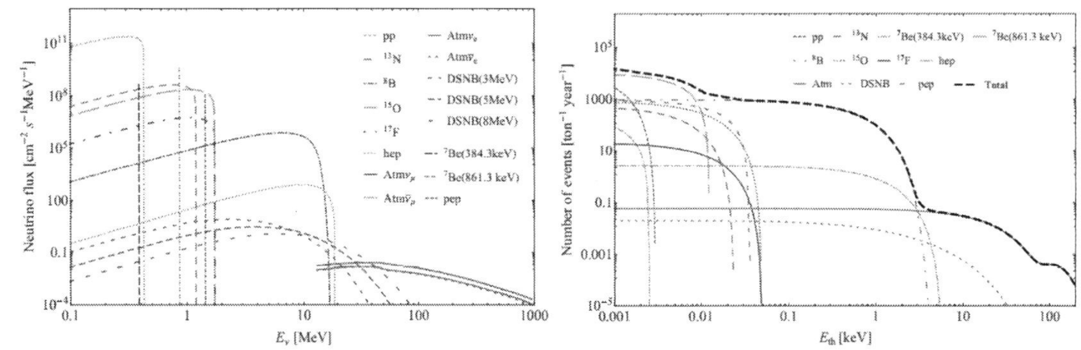

Figure 1. (Left panel) Neutrino fluxes on Earth relevant for direct dark matter detection. From low to high energies: solar, diffuse supernova and atmospheric neutrinos. (Right panel) Total number of neutrino background events per unit of ton-year as a function of energy threshold for a xenon target considering $E_{up} = 400$ keV (dashed black line); partial contributions relative to specific neutrino fluxes are also shown.

the kinematics requires a very low energy threshold to detect it. However, next generation DM experiments will have the potentiality to detect such a process limiting the sensitivity to WIMP discovery. This limit has been quantified by Billard et al [3] which have introduced the concept of *neutrino isoevent contour line*. This line is obtained by drawing on the WIMP-nucleon cross section vs WIMP mass parameters plane the best 90% C.L. exclusion limit achievable at each WIMP mass by a DM experiment for a one neutrino event exposure. This limit depends on the WIMP and neutrino rates which in turn depend on the dynamics of the interactions. In this work, we show the implication of some reasonable modifications to the standard differential cross section for a CENS process that have impact on the number of coherent neutrino background events seen by future DM detectors. In Section 2 we describe the neutrino flux at Earth and the standard formulae for the CENS cross section. In Section 3 we propose the modification of the Helm Form Factor (FF) for the CENS process while in Section 4 we explore the possibility of a partial coherency of the scattering process. Finally, in Section 5 we evaluate the impact on the event rate for both the proposed modifications and in Section 6 we draw our conclusions.

2. Neutrino flux and cross section

In this section we evaluate the neutrino background event rate starting from the description of the single neutrino flux components. The main neutrino fluxes at Earth can be divided into three categories: solar, atmospheric and diffuse supernovae neutrinos. Solar neutrinos are produced by the nuclear synthesis reactions inside the Sun such as the proton proton (*pp*) or Carbon-Nitrogen-Oxygen (CNO) chains. They are characterized by energies lower than ~ 2 MeV. The total flux produced in the *pp* chain is 5.94×10^{10} cm^{-2} s^{-1}. Another important contribution is represented by the Diffuse Supernova Neutrino Background (DSNB) generated by all the past supernova explosions in the Universe. The neutrino spectrum of a core-collapse supernova is assumed to be a Fermi-Dirac spectrum with temperatures in the range 3-8 MeV. DSNB flux covers energies up to hundreds of MeV. The last component is constituted by atmospheric neutrinos produced through cosmic ray collisions in the Earth atmosphere. The energy range of interest is approximately 10-1000 MeV. In Fig. 1 (left panel) all the neutrino flux components just mentioned are shown in detail.

As predicted in the SM neutrinos can scatter off coherently with a nucleus through a neutral current process. The coherence of the process is determined by the length scale of the interaction which in turn is related to the inverse of the momentum exchanged between neutrino and nucleus. The differential cross section depends on the neutrino energy E_ν and nuclear recoil energy E_r

$$\frac{d\sigma(E_\nu, E_r)}{dE_r} = \frac{G_f^2}{4\pi} Q_w^2 m_N \left(1 - \frac{m_N E_r}{2E_\nu^2}\right) |F(q^2)|^2 \,, \tag{1}$$

where G_f is the Fermi coupling constant, $Q_w = N - (1 - 4\sin^2\theta_w)Z$ is the weak hypercharge with N the number of neutrons, Z the atomic number and θ_w is the weak mixing angle. The nucleus mass is denoted by m_N while $|F(q^2)|^2$ is the nuclear form factor as a function of the momentum transfer q. From kinematics it can be shown that the maximum recoil energy and the minimum neutrino energy (in the limit $m_N \gg E_\nu$) are

$$E_r^{max} = \frac{2m_N E_\nu^2}{(m_N + E_\nu)^2} \quad \text{and} \quad E_\nu^{min} = \sqrt{\frac{m_N E_r}{2}} \,, \tag{2}$$

respectively. The neutrino differential rate and the integrated number of neutrino events are given respectively by

$$\frac{dR}{dE_r} = \eta \times \int_{E_\nu^{min}} \frac{dN}{dE_\nu} \times \frac{d\sigma(E_\nu, E_r)}{dE_r} dE_\nu \quad \text{and} \quad N_\nu(E_{th}) = \int_{E_{th}}^{E_{up}} \frac{dR}{dE_r} dE_r \,, \tag{3}$$

where η is the number of nucleus targets per unit of detector active mass and dN/dE_ν represents the total neutrino flux. The value of E_{up} represents the upper limit of the nucleus recoil energy. The number of neutrino events per unit of exposure as a function of the energy threshold E_{th} is shown in Fig. 1 (right panel).

3. Form factor modification

In Section 2 we have discussed the differential cross section for the ν-nucleus process. The form factor $|F(q^2)|^2$ introduced in Eq. 1 plays an important role. It takes into account the portion of the nucleus that contributes to the interaction. For this reason it is also connected with the coherency of the process. The result shown in Fig. 1 (right panel) is obtained considering the Helm parametrization of the form factor [4], following the prescription in [3]. This parametrization is certainly suitable for an interaction with equal couplings between the probe and all nucleons as assumed in the standard WIMP-nucleus interaction. However, the neutrino coupling is weighted by the Q_w^2 factor. Thus the contribution of ν-proton scattering is lower than the ν-neutron one because of the presence of the factor $(1 - 4\sin^2\theta_w) \simeq 0.076$. For this reason it would be better to use a form factor in which this feature is taken into account. Following [5] we propose a new parametrization in the form

$$F(q^2) = \frac{1}{Q_w}[NF_N(q^2) - Z(1 - 4\sin\theta_w{}^2)F_Z(q^2)]. \tag{4}$$

The nucleus form factor has been considered as a weighted combination between two different structure functions for the neutron and proton distributions alone, $F_N(q^2)$ and $F_Z(q^2)$. For these single structure functions we use the Helm parametrization described by the following expression

$$|F(q^2)_H|^2 = \left(\frac{3j_1(qR)}{qR}\right)^2 \exp\left[-q^2 s^2\right], \tag{5}$$

where j_1 is a spherical Bessel function of the first kind, R is the effective nuclear radius of the distribution and $s \simeq 0.9$ fm is the nuclear skin thickness. In this case we need to know the proton and the neutron radius, R_Z and R_N respectively. For R_Z we can use the data of the nuclear ground state charge radii [6] obtained by combining analyses of optical and absolute radii measured by muonic spectra and electronic scattering experiments. These techniques probe the root-mean-square (rms) nuclear radii of the proton distribution. For the case of xenon ($N = 77$ and $Z = 54$) we find that $R_Z \simeq 4.781$ fm.

Since the neutron radius is difficult to probe, with this technique we can estimate it using the following argument. Protons and neutrons are bound together inside a nucleus through strong interaction and so we expect $R_Z \propto (Z)^{1/3}$ and $R_N \propto (N)^{1/3}$. However, protons experience also the repulsive electromagnetic force, thus we expect the neutron radius to be smaller. We parametrize this difference adding a factor α. Putting all together we can write that

$$R_N = R_Z \cdot \alpha \cdot (N/Z)^{1/3}. \quad (6)$$

For α we assume a value of 0.9, which means that considering the same number of protons and neutrons the effect of the atomic repulsion would account for a decrease of the radius of the neutrons distribution of 90%. This value is in good agreement with the work of the PREX collaboration [7] in which for the first time the difference between R_N and R_Z of ^{208}Pb has been measured through an electroweak observation. In Fig. 2 (left panel) the comparison between the Helm form factor in Eq. 5 for xenon (red dashed line) and our modification in Eq. 4 for $\alpha = 0.9$ (solid blue line) is shown. The difference between the two parametrizations is clearly visible.

4. Coherency of the CENS process

In this section we discuss and evaluate the possibility for a transition towards decoherency in a CENS process. The problem has also been studied recently by the TEXONO collaboration [8]. Assuming that the degree of coherency could be quantified by a measurable parameter $\beta \in [0, 1]$ they obtained that the ratio between the neutrino cross section with a nucleus with Z protons and N neutrons ($\sigma_{\nu-N}$) and the cross section of a single neutron ($\sigma_{\nu-n}$) can be written as

$$\frac{\sigma_{\nu-N}}{\sigma_{\nu-n}} = \epsilon^2 Z + N + Z(Z-1)\epsilon^2\beta + N(N-1)\beta - 2ZN\epsilon\beta, \quad (7)$$

where $\epsilon \equiv 1 - 4\sin\theta_w{}^2$. For $\beta = 1$ the expression in Eq. 7 becomes $\propto [\epsilon Z - N]^2$, which represents the full coherency regime, while $\beta = 0$ implies the opposite condition of total decoherency for which $\propto [\epsilon^2 Z + N]$. Notice that in the first case the ratio in Eq. 7 is precisely equal to the weak hypercharge to the square, Q_w^2.

With these definitions in mind it is possible to evaluate the total cross section, $\sigma(E_\nu)$, as a function of the neutrino energy integrating Eq. 1 between zero and E_r^{max}. In Fig. 2 (right panel) we show the total cross section evaluated combining our modifications to the form factor and to the coherency hypothesis. With the dashed red line it is shown the total cross section (Helm plus full coherency) as used in the Billard et al. paper [3], while with the solid blue line it is shown the prediction obtained when using the form factor presented in Eq. 4 with $\alpha = 0.9$. The difference is not so dramatic, however for a neutrino energy in the range 10-100 MeV the inclusion of the proper form factor increases the cross section prediction by a factor 1.007-1.3 respectively. On the contrary, the coherency correction gives a more pronounced effect, lowering the cross section by a flat factor of about 68.7, going from $\beta = 1$ (full coherency) to $\beta = 0$ (no coherency).

5. CENS event rate

We now evaluate the impact of our modifications to the differential neutrino event rate in Eq. 3. In Fig. 3 (left panel) it is shown the comparison between the differential number of neutrino background events as a function of the recoil energy per ton-year of a DM xenon experiment for different choices of β and α. Since the coherency correction does not depends on the energy (see

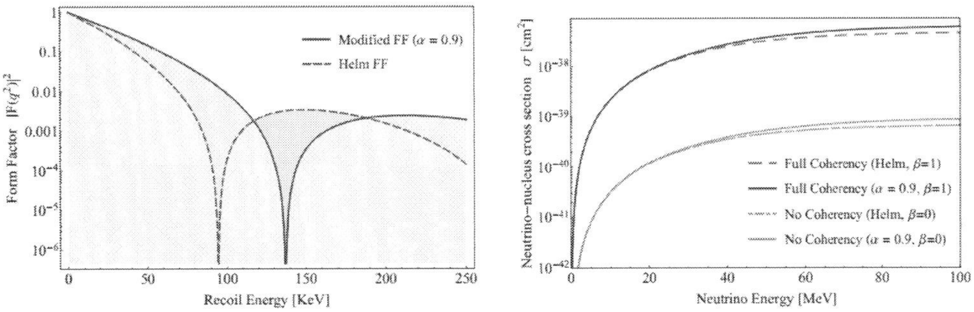

Figure 2. (Left panel) Comparison between the standard parametrization of the Helm form factor, $|F(q^2)_H|^2$, for a xenon nucleus with dashed red line, and our parametrization of Eq. 4 for $\alpha = 0.9$ with the solid blue line as a function of the nucleus recoil energy in keV. (Right panel) Integrated cross section for our proposed changes.

Eq. 7), this results in a flat decrease of the differential number of events by a factor ~ 68.7, which lowers remarkably the number of background with the respect to the standard formula ($\beta = 1$). For $1 > \beta > 0$ all the possible correction factors between 1 and 68.7 are in principle possible. In the case of the form factor modification the correction is energy dependent and it is less pronounced than the coherency correction. In order to quantify the impact of the implemented correction is better to look at the integrated number of neutrino background events in a given experimental energy window. Driven by a pragmatic motivation, we consider the ratio between the integrated number of neutrino events considering the Helm form factor and the our form factor implementation

$$R(E_{th}) \equiv \int_{E_{th}}^{E_{up}} \frac{(dR^{Helm})}{dE_r} dE_r \Big/ \int_{E_{th}}^{E_{up}} \frac{(dR^{FF})}{dE_r} dE_r \ , \tag{8}$$

and we choose some typical values of E_{th} and E_{up} for currently ongoing and future xenon experiments such us XENON1T/XENONnT [9], LZ [10] and DARWIN [11]. In Fig. 3 (right panel) it is shown the ratio in Eq. 8 having fixed E_{up} to 40, 50 and 60 keV, as a function of the threshold energy E_{th} varying from 2 to 8 keV. The different points represent the particular choice of XENON1T (green triangle), XENONnT (blue cross), LZ (black star) and DARWIN (red disk) for which R has the value of 0.77, 0.78, 0.87 and 0.78 respectively. This means that the relative correction to the number of neutrino background events is between 15-30% for the most used value of E_{up} and E_{th}. For example, in the case of DARWIN which aims to reach an exposure of 200 tonne year, the integrated number of neutrino background events will be 12.8 instead of 10. This would result in a significant change of the sensitivity or of the discovery limit of WIMPs. Indeed, since the neutrino and WIMP spectra match relatively well, in the case of a significant neutrino contribution it can be shown [3] that the WIMP discovery limit goes like $\sigma_{CL} \propto \sqrt{(1+\xi^2 N_\nu)}/N_\nu$, where ξ is the systematic error in percentage on the neutrino contribution. Considering that the neutrino flux uncertainty of the two neutrino components of interest, namely atmospheric and DSNB, is quite big (20% and 50% respectively), it is of vital importance to reduce as much as possible the uncertainties on the cross section inputs, the uncertanties related to the neutron distribution radius and to find the right coherence dependence with a dedicated experimental campaign.

6. Conclusions

In this work we discuss how modifications to the neutrino-nucleus form factor and the coherency hypothesis of CENS process could lead to sizeable change of the neutrino irreducible background.

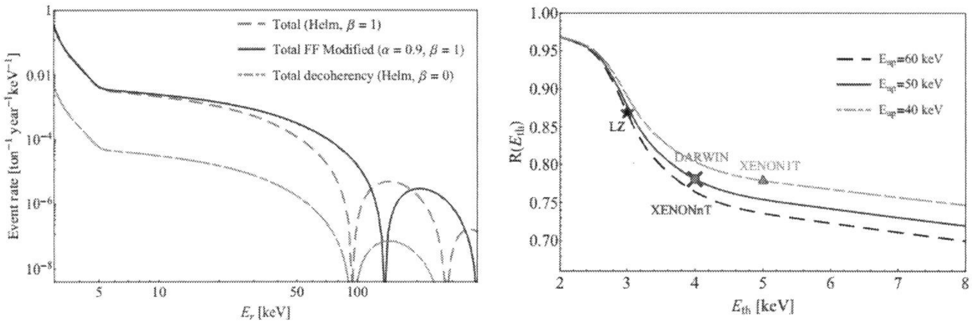

Figure 3. (Left panel) Neutrino event rate as a function of the recoil energy E_r. The dashed red line represents the standard prediction, the solid blue line is the prediction with the inclusion of the proper form factor ($\beta = 1$), while the dash-dotted green line represents the differential rate for total decoherency regime ($\beta = 0$) and with a Helm form factor. (Right panel) Ratio between the integrated number of neutrino events considering the standard cross section and the form factor correction as a function of the threshold energy (see Eq. 8). We fix E_{up} to 40, 50 and 60 keV, and we show the ratio as a function of the threshold energy E_{th}. The different points represent the particular choice of E_{th} made by XENON1T (green triangle), XENONnT (blue cross), LZ (black star) and DARWIN (red disk).

We show, first of all, that the adoption of the Helm form factor, as done in [3], is not suitable for the case of CENS. Neutrinos couple preferentially with neutrons, therefore probing more the neutron distribution than the proton one. We propose to use a form factor obtained by a weighted combination between the two structure functions of the neutron and of the proton, $F_N(q^2)$ and $F_Z(q^2)$. Since $F_N(q^2)$ has not been precisely determined yet we propose a simple parametrization to estimate it. We thus evaluate the differential neutrino spectrum and the total number of neutrino events. We have shown that the usage of our form factor increases the number of background neutrino events by a factor 15-30% in the case of future xenon DM experiments like XENON1T, XENONnT, LZ and DARWIN. We consider also the possibility for the CENS process not to be completely coherent including a coherency correction factor β. This correction gives a more substantial effect on the predicted number of neutrino events lowering it down to a factor ~ 68.7 in the case of total decoherence ($\beta = 1$). All these corrections to the CENS cross section prediction substantially change the neutrino isoevent contour line [12]. Since the WIMP discovery limit depends crucially on the uncertainty in the neutrino background prediction, it is of outmost importance to perform a dedicated measure of the CENS.

References

[1] Akerib D S *et al* 2016 *Phys. Rev. Lett.* **116** 161301
[2] Freedman D Z 1974 *Phys. Rev.* D **9** 1389
[3] Billard J *et al* 2014 *Phys. Rev.* D **89** 023524
[4] Helm R 1956 *Phys. Rev.* **104** 1466
[5] Amanik P S and McLaughlin G C 2009 *J. Phys.* G **36** 015105
[6] Angeli I and Marinova K P 2013 *Atomic Data and Nuclear Data Tables* **99** Pages 69-95
[7] Hagen G *et al* 2016 *Nature Physics* **12**, 186190
[8] Kerman S 2016 *Phys. Rev.* D **93** 113006
[9] Aprile E *et al* XENON Collaboration 2016 JCAP04 027
[10] Akerib D S *et al* LUX-ZEPLIN Collaboration 2015 arXiv:1509.02910 [physics.ins-det]
[11] Aalbers J *et al* 2016 JCAP 1611 no.11, 017
[12] Cadeddu M and Picciau E 2017 (in preparation)

Measurement of R(D^*) with hadronic τ decays

B G Siddi[1,2] on behalf of the LHCb collaboration

[1]Università degli studi di Ferrara, Dipartimento di fisica, Via Saragat 1, 44122, Ferrara
[2]CERN, CH-1211 Geneva 23, Switzerland

E-mail: bsiddi@cern.ch, siddi@fe.infn.it

Abstract. Lepton universality, described in the Standard Model (SM), predicts equal coupling between gauge bosons and the three lepton families. SM extensions give additional interactions, implying in some cases a stronger coupling with the third generation of leptons. Semileptonic decays of b-hadrons provide a sensitive probe to such New Physics effects. The presence of additional charged Higgs bosons, required by such SM extensions, can have significant effect on the semileptonic decay rate of $B^0 \to D^*\tau\nu$. A probe of new physics effects is the measurement of the quantities: $R(D^*) = \frac{\mathcal{B}(\bar{B}^0 \to D^{*+}\tau^-\bar{\nu}_\tau)}{\mathcal{B}(\bar{B}^0 \to D^{*+}\mu^-\bar{\nu}_\mu)}$ and $R(D) = \frac{\mathcal{B}(\bar{B}^0 \to D^{+}\tau^-\bar{\nu}_\tau)}{\mathcal{B}(\bar{B}^0 \to D^{+}\mu^-\bar{\nu}_\mu)}$. The combination of experimental measurements performed by BaBar, Belle and LHCb observing the channel where the τ decays in leptons, gives a deviation from the SM prediction of about 4 σ. It is therefore important to perform additional measurements in this sector in order to improve the precision and confirm or disprove this deviation. Another possibility is to perform this measurement using the channel where the semileptonic τ decays in 3 pions. This in LHCb allows to have e better reconstruction of vertices and other kinematic variables. Results obtained by LHCb on $B^0 \to D^*\tau\nu$ decays, where the τ decays hadronically, are reported.

1. The LHCb Detector

The LHCb detector [1] (see Fig. 1) is a single arm spectrometer covering the rapidity region $2 < \eta < 5$. It is optimized to study hadron decays containing b and c quarks. These are important to analize CP violation, rare decays, and heavy quark production. It has an excellent vertex resolution for B and D hadrons, a good momentum and invariant mass resolution. The detector provides good Particle Identification (PID) capabilities; the kaon identification efficiency is 88% and the pion mis-indentification is 3%. LHCb during Run 1, collected about $1.0 fb^{-1}$ of data at $\sqrt{s} = 7$ TeV and about 2.0 fb^{-1} at $\sqrt{s} = 8$ TeV.

2. B hadron semileptonic decays with τ leptons in final states

Semileptonic decays of b hadrons are successfully studied in B-factories with high purity and high statistics $D^{(*)}\tau\nu$ samples; despite the hadronic environment LHCb is also able to study such kind of decays and extend to other b hadrons thanks to their high boost and excellent vertexing of the detector. However there are some analysis challenges in an hadronic enviroment, e.g., it needs to find the correct kinematic variables in order to distinguish between signal and background, suppress the background due to additional charged and/or neutral particles, and find the best normalization channel. These challenges have different levels of importance and difficulty, and different solution between analyses, expecially between muonic and hadronic τ decays.

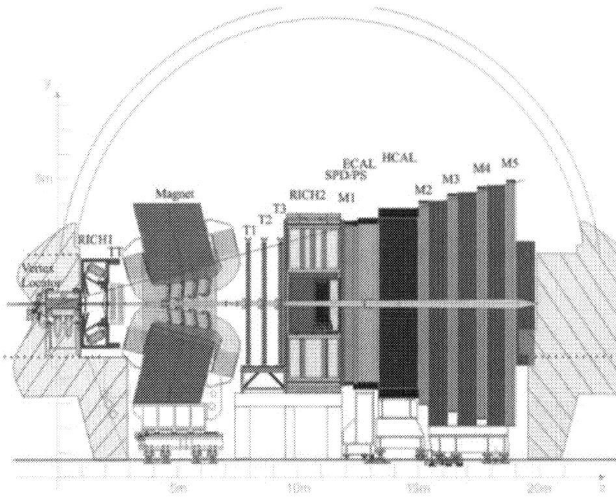

Figure 1. Schematic view of the LHCb Detector.

3. τ leptons with hadronic final state

A way to measure R(D^*) is to reconstruct the τ^- in the hadronic $\pi^-\pi^+\pi^-\nu_\tau$ decay channel. This is a semileptonic decay without charged leptons in the final states. In this case what is measured is:

$$K(D^*) = \frac{\mathcal{B}(B^0 \to D^*\tau\nu)}{\mathcal{B}(B^0 \to D^*\pi\pi\pi)} \tag{1}$$

The two decay modes shares the same final state and in this case most of the systematic uncertainties will cancel. To compute $R(D^*)$ a second equation will be needed:

$$R(D^*) = K(D^*) \times \frac{\mathcal{B}(B^0 \to D^*3\pi)}{\mathcal{B}(B^0 \to D^*\mu\nu)} \tag{2}$$

The quantities multiplied by $K(D*)$ are then taken as external inputs.

4. Background

4.1. Vertex inversion

The most abundant background source is due to hadronic B decays into $D^*3\pi X$. Such background has a branching fraction that is about 100 times the signal one.

$$\frac{\mathcal{B}(B^0 \to D^*3\pi X)}{\mathcal{B}(B^0 \to D^*\tau\nu)_{SM}} \sim 100 \tag{3}$$

In Fig. 2 a schematic view of the $B^0 \to D^*3\pi X$ decay is presented.

Thanks to the good precision on the τ vertex reconstruction, it is possible to apply the so called vertex inversion cut. This consists on the requirement that the τ vertex is downstream with respect to the B^0 one with a significance of at least 4σ. A schematic view of this requirement is shown in Fig. 3. In this way the background coming from these decays is suppressed by about 3 orders of magnitude.

The distribution of the quantity Δ_z/σ for different background sources is presented in Fig. 4. The $D^*3\pi$ contribution (grey) is basically suppresed, after a value of $\Delta_z/\sigma = 4$.

IOP Conf. Series: Journal of Physics: Conf. Series **956** (2017) 012015 doi:10.1088/1742-6596/956/1/012015

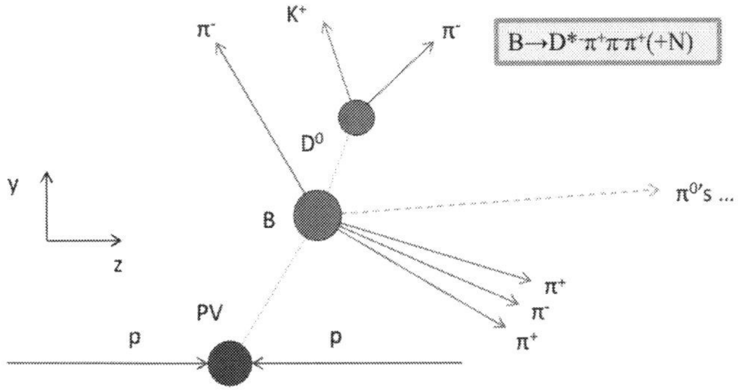

Figure 2. Schematic view of the $B^0 \to D^* 3\pi X$ decay.

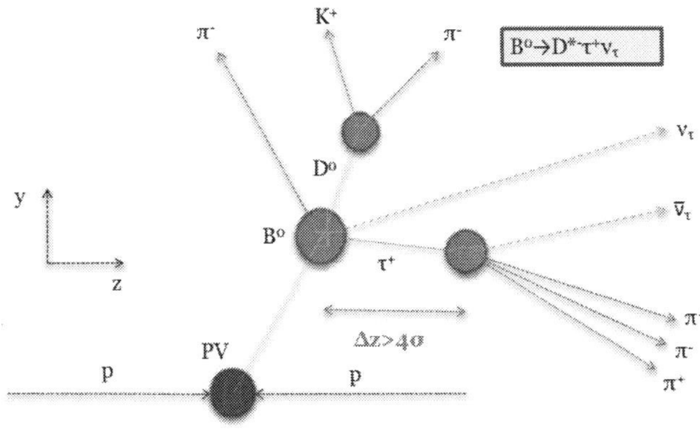

Figure 3. Schematic view of the vertex cut requirement.

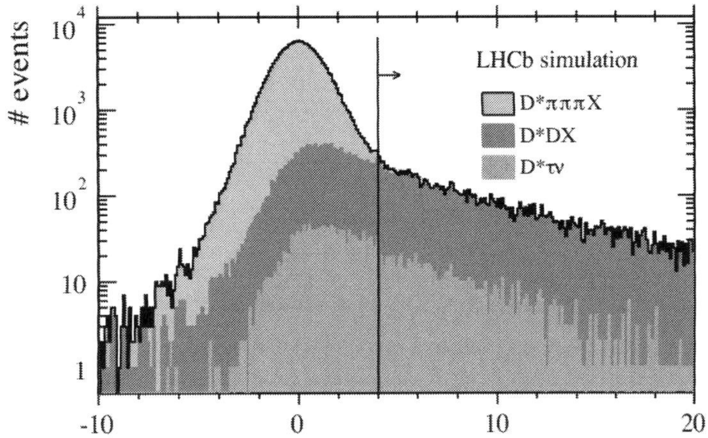

Figure 4. Distribution of the vertex significance (Δ_z/σ) for different background sources. The black line represents the applied cut.

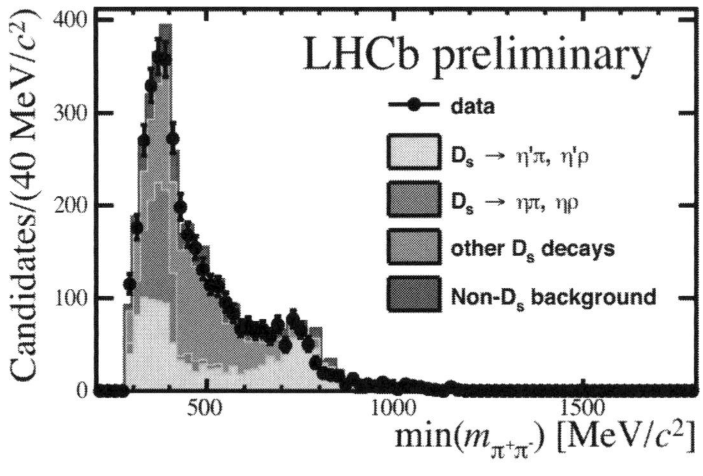

Figure 5. D_s model fit on data. The different contributions are shown.

4.2. Double charm decays

The remaining background consists of B^0 decays where the 3π vertex is transported away from the B^0 one by a charm carrier, such as D^0, D^+ and D_s. This background has a yield that is about 10 times the signal:

$$\frac{\mathcal{B}(B^0 \to D^* D_{(s)}^{(*)}; D_{(s)}^{(*)} \to 3\pi + X)}{\mathcal{B}(B^0 \to D^* \tau\nu)} \sim 10 \qquad (4)$$

To limit these background sources, LHCb has three very good tools: the 3π dynamics; the isolation criteria against charged tracks and neutral energy deposits; partial reconstruction techniques in the signal and background hypotheses. All these variables are used to train a Boosted Decision Tree (BDT) in order to discriminate double charm decays from the signal.

4.3. D_s decay model

The D_s decay model has been determined directly from data, using a sample obtained using a BDT output that is enriched in such decays (high purity is reached). In order to retrieve the sub-decays relative fractions, a simultaneous fit in the minimum and maximum of the opposite sign 2π mass, in the same sign pions ($\pi^+\pi^+$) mass and 3π mass has been performed. The result of the fit is shown in Fig. 5. The total PDF of the fit contains:

- D_s decays where at least 1 pion is from η or η': $\eta\pi$, $\eta\rho$, $\eta'\pi$ and $\eta'\rho$;
- D_s decays where at least 1 pion is from an IS (Intermediate Resonance) other than η, η': ISπ, ISρ (IS could be ω, ϕ);
- D_s decays where none of the 3 pions comes from an IS, subdivided in: $K^0 3\pi$, $\eta 3\pi$, $\eta' 3\pi$, $\omega 3\pi$, $\phi 3\pi$, 3π non resonant final state.

The weights obtained by the fit are then used to construct the D_s template utilized in the final fit.

5. Normalization

The normalization channel has to be as similar as possible to the signal channel in order to cancel all systematics linked to trigger selection, particle identification and selection cuts. Due

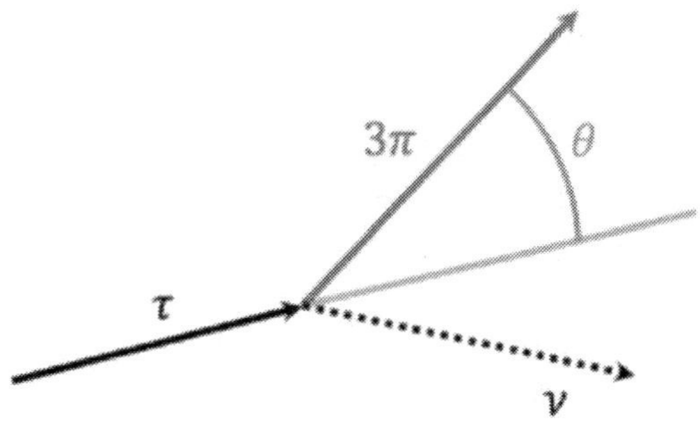

Figure 6. Schematic view of θ, the angle between τ and 3π direction.

to this, in this analysis what is measured is the ratio:

$$K(D^*) = \frac{\mathcal{B}(B^0 \to D^*\tau\nu)}{\mathcal{B}(B^0 \to D^*\pi\pi\pi)} \tag{5}$$

The two decays differ by the presence of a softer pion and D^* in the signal decay due to the two extra neutrinos, and the kinematics of the 3π system. This will give a residual effect on the efficiency ratio.

6. Signal extraction

6.1. Signal reconstruction

The presence of the neutrino in the final state of the signal decay can be inferred, up to a two-fold ambiguity, by exploiting the flight direction of the τ. In this way it is possible to compute the τ momentum with the following equation:

$$|\vec{p}_\tau| = \frac{(m_{3\pi}^2 + m_\tau^2)|\vec{p}_{3\pi}|\cos\theta \pm E_{3\pi}\sqrt{(m_\tau^2 - m_{3\pi}^2)^2 - 4m_\tau^2|\vec{p}_{3\pi}|^2\sin^2\theta}}{2(E_{3\pi}^2 - |\vec{p}_{3\pi}|^2\cos^2\theta)} \tag{6}$$

where θ is the angle between τ and 3π direction as shown in Fig. 6.

The ambiguity can be resolved by choosing the maximum value for the opening angle between the three charged pions system and the direction of the τ lepton, as in the following equation:

$$\theta_{max} = \arcsin\left(\frac{m_\tau^2 - m_{3\pi}^2}{2m_\tau|\vec{p}_{3\pi}|}\right) \tag{7}$$

This will introduce a negligible bias in the signal reconstruction.

6.2. Fit

An extended maximum likelihood 3-dimensional fit, using templates has been performed. The fitted variables are:

- q^2, the squared momentum tranferred to the $\tau - \nu$ system;
- τ decay time;
- The output of the BDT extracted from simulated and data-driven control samples.

The fit model used to extract the signal yield consists of 5 main categories:

- Signal described by the sum of two τ decays, $\tau \to \pi\pi\pi\nu$ and $\tau \to \pi\pi\pi\pi^0\nu$;

- $B^0 \to D^{**}\tau\nu$ decays;
- Double charm components;
- $B^0 \to D^*3\pi X$ decays;
- Combinatorial background.

7. Systematic uncertainties

The main systematic uncertainty is due to the size of samples used for the construction of the fitted templates. Another source of main systematics is the external branching ratio of the $B^0 \to D*3\pi$ decay. This value was known in the particle data group 2014 with a 11% precision. Thanks to the last BaBar measurement of this decay, this systematics has been reduced to 4.3% [2]. In the uncertainty computation, another contribution is given by the knowledge of the $B \to D^*(D_s, D^0, D^+)X$ and $B^0 \to D^*3\pi X$ background shapes. The size the total systematics is above the statistical precision but there is room for improvements, i.e., with more data and simulated samples, and a better knowledge of the external branching franctions using also results from other experiments.

8. Conclusion

Semitauonic B decays are a great tool to discover new physics. There are several measurement of R in LHCb, e.g. $R(D^*)$, $R(D)$, $R(J/\psi)$, $R(\Lambda_c)$ and more modes are possible. Thanks to the LHCb excellent performances, it is possible to reconstruct hadronic τ decays with a good precision. This allowed to extract the signal for hadronic $R(D^*)$ with a 7% of precision using just the Run1 data. This measurement is competitive with the LHCb one in which the muonic τ decay has been used, and with the world average [3]. However, it is possible to use such decays to study not only R but also other quantities such as angles, polarizations, form factors.

References

[1] Aaij R *et al* 2014 LHCb Detector Performance CERN-LHCB-DP-2014-002
[2] Lees J P *et al.* 2016 Measurement of the $B^0 \to D^{*-}\pi^+\pi^-\pi^+$ branching fraction *Phys.Rev.* D**94** 091101
[3] Heavy Flavor Averaging Group (HFLAV) Averages of b-hadron, c-hadron, and τ-lepton properties as of summer 2016 arXiv:1612.07233

Study on TPB as wavelength shifter for the new ICARUS T600 light collection system in the SBN program.

M Spanu[1], A Falcone[1], R Mazza[2], A Menegolli[1], M C Prata[1], G L Raselli[1], M Rossella[1], M Torti[1].

[1] University of Pavia - INFN Pavia, Via Bassi 6, Pavia (ITALY)

[2] INFN Milano Bicocca, Piazza dell'Ateneo Nuovo, 1, Milan (ITALY)

E-mail: maura.spanu@pv.infn.it

Abstract.
In the last 30 years, the incredible experimental progress made in the studies of neutrino oscillation allowed to better understand the pattern of neutrino masses and neutrinos mixing. However, further investigation are necessary, in particular concerning a series of experimental anomalies, observed in different neutrino experiments, which are uncorrelated with each other but all hinting at oscillation phenomena. The goal of the new Short Baseline Neutrino program is to perform sensitive searches for ν_e appearance and ν_μ disappearance in the Booster Neutrino Beam in order to understand experimental anomalies in neutrino physics and to perform the most sensitive search for sterile neutrinos at the eV mass-scale. The experiment includes three Liquid Argon Time Projection Chamber detectors located along the Booster Neutrino Beam line at Fermilab. In this paper, the functioning of the Short Baseline Neutrino far detector, ICARUS-T600, is shown. In particular, this work is focused on the detector light collection system and on its upgrade concerning the wavelength shifting of the liquid argon scintillation from vacuum ultra-violet into visible light.

1. Introduction

The experimental observations of neutrino oscillations have established a picture consistent with the mixing of three neutrino flavours ν_e, ν_μ and ν_τ with three mass eigenstates ν_1, ν_2 and ν_3 whose mass differences turn out to be relatively small. However, in recent years, several experimental anomalies have been reported [1]. A possible phenomenological model with light sterile neutrinos is widely used in order to explain the anomalies in neutrino oscillation pattern described above: the so called $3+1$ model. According to this model, the existing neutrino states are the three interacting lepton flavours and a fourth non-interacting *sterile* neutrino state. Global analysis of all the data relevant for the sterile neutrino hypothesis were performed. The results of the $3+1$ global fit are reported in [2].

2. The SBN program

The Short-Baseline Neutrino (SBN) physics program [3], emerging from a European - US collaboration, will deliver a rich and compelling physics opportunity, including the possibility to investigate the experimental anomalies in neutrino physics and to perform the most sensitive search to date for sterile neutrinos at the eV mass-scale through both appearance

IOP Conf. Series: Journal of Physics: Conf. Series **956** (2017) 012016 doi:10.1088/1742-6596/956/1/012016

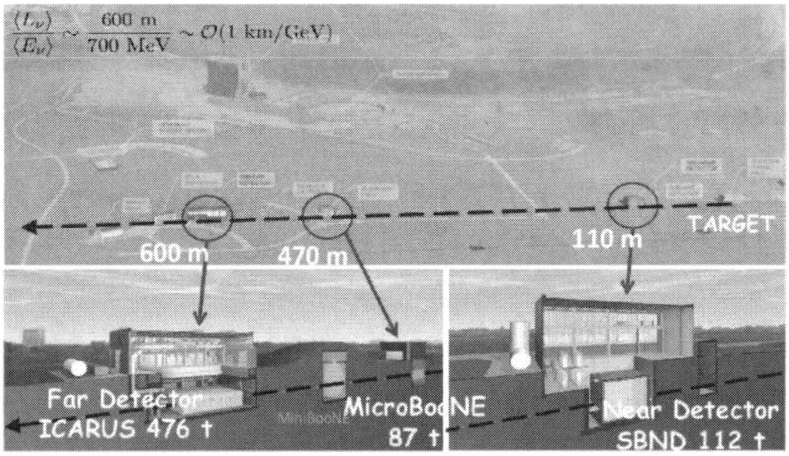

Figure 1. Map of the Fermilab neutrino beamline area showing the axis of the BNB (blue dashed line) and locations of the SBN detectors at 110 m, 470 m and 600 m.

and disappearance oscillation channels. The SBN program includes three Liquid Argon Time Projection Chamber (LAr-TPC) detectors (Fig. 1) namely the Short-Baseline Near Detector (SBND), Micro Booster Neutrino Experiment (MicroBooNE), and Imaging Cosmic And Rare Underground Signals (ICARUS) experiments, located on-axis along the Booster Neutrino Beam (BNB) line at Fermilab. The size and position of the three detectors allow to characterize the beam before oscillations in the near detector and to measure simultaneously ν_e appearance and ν_μ disappearance in the MicroBooNE and ICARUS detectors.

3. The ICARUS T600 Time Projection Chamber

ICARUS T600 will act as the SBN far detector. It successfully operated at Gran Sasso National Laboratories (LNGS) on the CERN Neutrino to Gran Sasso (CNGS) muon neutrino beam from 2010 to 2012 and it was the first large scale LAr TPC to be exposed to a neutrino beam and the largest existing LAr TPC for neutrino physics [4].

It consists of a large cryostat split into two identical modules and filled with 760 tons of ultra-pure liquid argon. Each module houses two TPCs separated by a common cathode, with a drift length of 1.5 m. Ionization electrons, abundantly produced by charged particles along their path (6000 electrons/mm for minimum ionizing particles at an electric field $E_D = 500\ V/cm$), are drifted under uniform electric field towards the TPC anode made of three parallel wire planes. A total of about 53000 wires are deployed, with 3 mm pitch, oriented on each plane at a different angles $(0°, +60°, -60°)$ with respect to the horizontal direction.

By appropriate voltage biasing, the first two wire planes (Induction-1 and Induction-2) provide signals in a non-destructive way, whereas the ionization charge is collected and measured on the last plane (Collection).

3.1. Geometrical reconstruction of an ionizing event in ICARUS T600

Charge signals on the TPC wire planes, from ionization electrons, allow to obtain three bi-dimensional pictures of the ionizing event in the plane normal to the drift direction [5]. The relative time of each ionization signal, combined with the electron drift velocity information ($v_D \sim 1.6\ mm/\mu s$), provides the position of the track along the drift coordinate. A three-dimensional image of the events can then be reconstructed by combining the wire coordinates on each plane and the measured drift time, with a resolution of about 1 mm^3 (Fig. 2). The precise

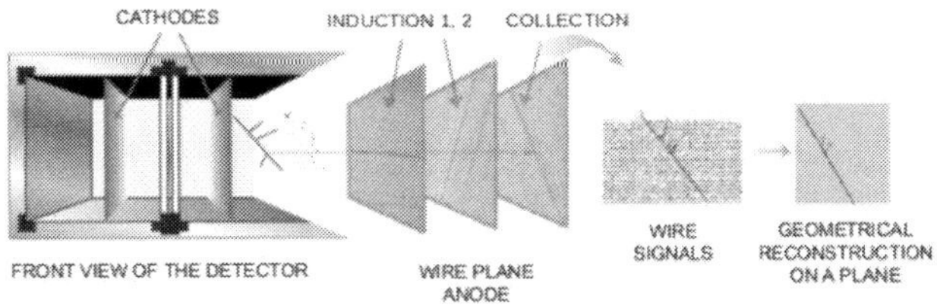

Figure 2. A charged particle ionization path in LAr and its geometrical reconstruction.

determination of the coordinate along the drift time direction, is derived from the absolute time of the ionizing event given by the detection of the scintillation light ($\sim$ 5000 VUV photons/mm at $E_D = 500\ V/cm$ with $\lambda \approx 128\ nm$) produced by LAr. The emission is due to the radiative decay of excited molecules Ar_2^* and it is characterized by two different decay times ($\sim 6\ ns$ and $\sim 1.5\ \mu s$) [6]. To detect the scintillation light, arrays of Photo Multiplier Tubes (PMT) are installed behind the wire planes.

3.2. The T600 light collection system

The future operation of the ICARUS T600 at shallow depth at Fermilab requires an improvement of the light collection system due to the higher cosmic muons rate background at shallow depth, in order to detect with full efficiency the prompt scintillation light from events with energy depositions down to $\sim 100\ MeV$ or even below. Moreover, the ICARUS light collection system, such as in most LAr detectors, is essential for generating a light-based trigger signal and identifying the time of occurrence of each interacting particle with high temporal precision.
The new T600 light collection system consists of 90×8" PMTs for each TPC, involving a total of 360 PMTs. The photomultiplier tube model and placement have been selected in order to satisfy the main light collection system requirement as an high detection coverage and granularity and a fast response [7]. However, a further requirement has been considered. The PMT photocathode is sensitive to $\lambda = 400 \div 420\ nm$ wavelength light whereas the LAr scintillation light is emitted at $\lambda = 128\ nm$. Therefore, the most efficient way to detect the scintillation light is to shift them to longer wavelengths where the PMT photocathode is sensitive. *Tetra-Phenyl-Butadiene* (TPB) is the most common wavelength shifter used in LAr experiments. It is a fluorescent organic chemical compound which radiates blue light with an emission spectrum peak at 430 nm wavelength. Generally, a thin layer of TPB is used to cover the PMT front surfaces or some internal sections of the detector.

4. The TPB as wavelength shifter: the evaporation system

A thin layer of TPB was deposited on the sensitive glass window of all the PMTs by a thermal evaporator available at CERN. It consists in a vacuum chamber housing two copper crucibles (Knudsen cells) connected to an external heating system which allows to increase their internal temperature. In high vacuum regime (few $10^{-5}\ mbar$), TPB starts to evaporate at $T \approx 190°C$. Through a small hole in the crucible lid, the TPB vapour diffuses in the vacuum chamber forming a thin film on the PMT surface, which is at room temperature. The evaporation system has been realized to be able to control the main evaporation process parameters. In particular, a thickness control system, consisting of a quartz crystal placed inside the vacuum chamber, an external oscillator and a thickness monitor, was used. The variation of the crystal oscillation frequency allows to measures the quantity (thickness) of material deposited on its surface. The monitor displays the coated thickness value both per time unit ($\mathring{A}/s$) and integrated on the whole process. Moreover, in order to obtain a good thickness uniformity, the PMT was fixed

IOP Conf. Series: Journal of Physics: Conf. Series **956** (2017) 012016 doi:10.1088/1742-6596/956/1/012016

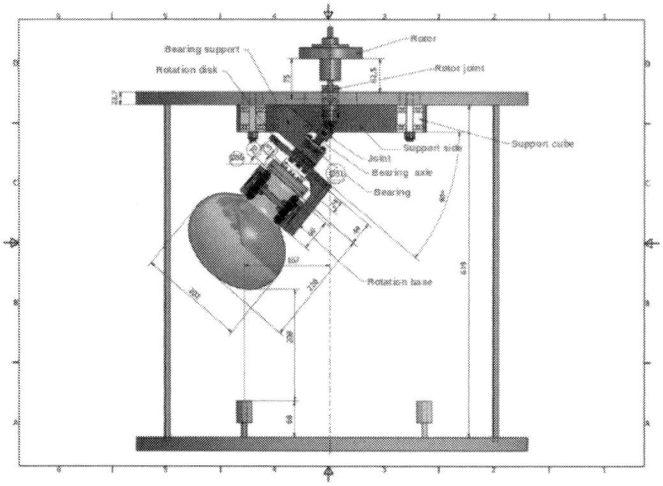

Figure 3. Schematic design of the evaporating chamber and PMT rotative support. The Knudsen cells are also shown in their location inside the vacuum chamber.

looking downwards with respect to the vertical direction by a dedicated rotating support fastened below the lid of the vacuum chamber (Fig. 3).

5. Quantum Efficiency measurements

The TPB coating uniformity is evaluated in terms of light yield conversion. For this reason, an optical device was set up allowing to measure the overall PMT and TPB Quantum Efficiency (QE), namely the ratio of the number of photoelectrons released in the photocathode to the number of photons absorbed by the wavelength shifter. Quantum efficiency is derived by the equation:

$$QE_{DUT}(\lambda) = QE_{NIST}(\lambda) \times \frac{I_{DUT}(\lambda) - I_{DUTdark}}{I_{NIST}(\lambda) - I_{NISTdark}} \tag{1}$$

where QE_{DUT} and QE_{NIST} are respectively the PMT under test and the calibrated photodiode (tabulated value) quantum efficiency, I_{DUT} and I_{NIST} are respectively the PMT and photodiode cathodic current values and I_{dark} are the dark currents.

5.1. The optical system

The optical device used for the QE measurements is shown in Fig. 4. The light source is a Deuterium lamp whose emission spectrum has two peaks at 121 nm and 161 nm. The focusing elbow allows sending the deuterium lamp light to the Monochromator. Focusing is made by a UV-reflective mirror. The light generated by the deuterium lamp light gets into the monochromator through a first slit, which allows to calibrate its intensity. A diffraction grating (120 $ridges/mm$) and a second slit allow to select the light wavelength and to regulate its range amplitude. A collimation system, consisting of two different collimator, allows to stabilizes the light beam size and to reduce its divergence. The light beam is sent towards a rotative mirror which allows to selectively direct it towards the PMT or the calibrated photodiode. The cathodic current from both the photomultiplier and the photodiode is alternately measured by a pico-ammeter.

IOP Conf. Series: Journal of Physics: Conf. Series **956** (2017) 012016 doi:10.1088/1742-6596/956/1/012016

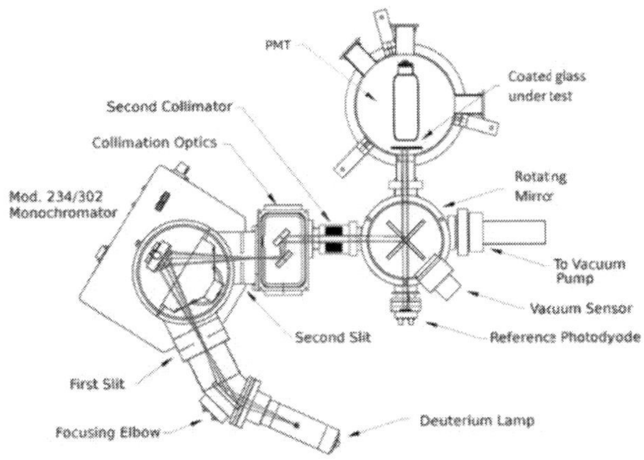

Figure 4. Experimental set-up of the optical device used for the Quantum Efficiency measurements.

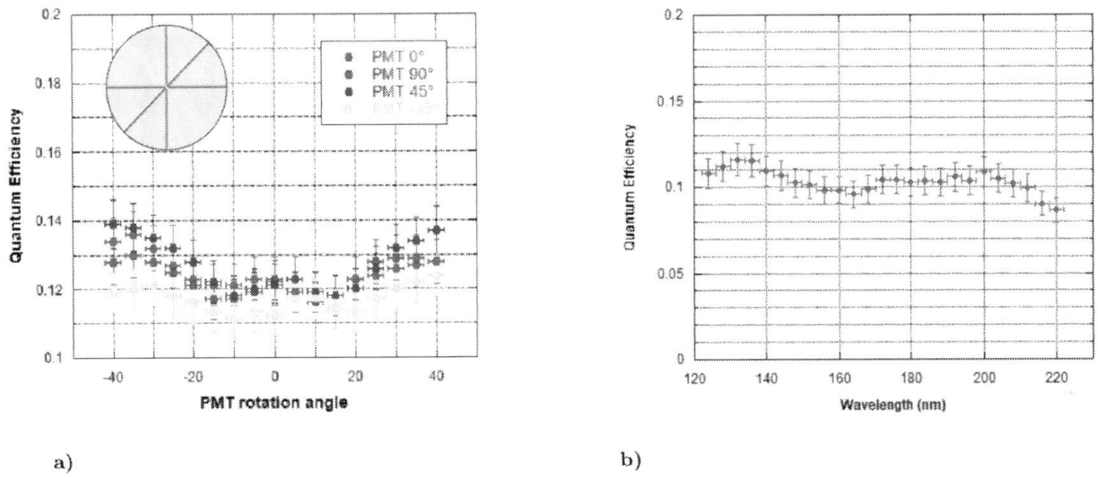

Figure 5. Results of the uniformity test on a photomultiplier. In *a)* the resulting overall Quantum Efficiency is shown on 4 different axis on the PMT surface. In *b)* the PMT response spectrum as a function of the incident light wavelength is also shown.

5.2. Measurements and results

The coating uniformity on the PMT sensitive window has been evaluated in terms of light yield conversion. To this purpose, the PMT was fixed inside the measuring chamber by means of a specific support ring, which allows to rotate the photomultiplier perpendicularly to its own axis. At first, the PMT response on its window center at $\lambda = 128\ nm$ was measured. After that, PMT was rotated in the horizontal plane with an angle between $-40°$ and $+40°$ with respect to the optical beam axis, in steps of $5°$. In this way, the PMT response was measured along a radius of the photomultiplier window. This measure was repeated after rotating the PMT around the optical axis respectively of $90°$, $45°$ and $-45°$. Results are shown in Fig. 5. The PMT response spectrum as a function of the incident light wavelength is also shown.

Test was repeated comparing the Quantum Efficiency response of other PMTs along one circle, only. Results, also showing a mainly flat response, are displayed in Fig. 6.

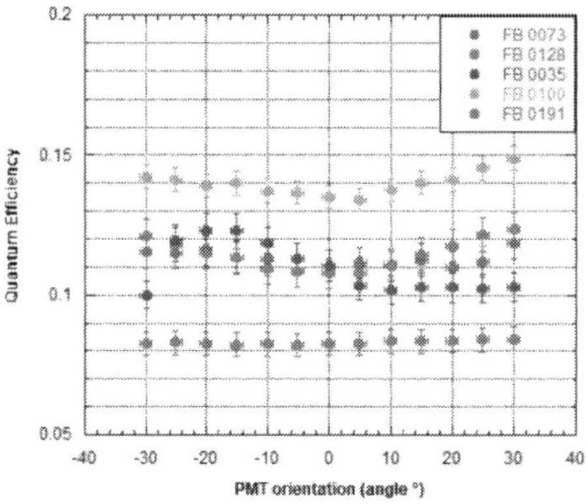

Figure 6. Results of the uniformity test on 5 PMTs. The resulting QE is shown ranging from 0.05 to 0.2.

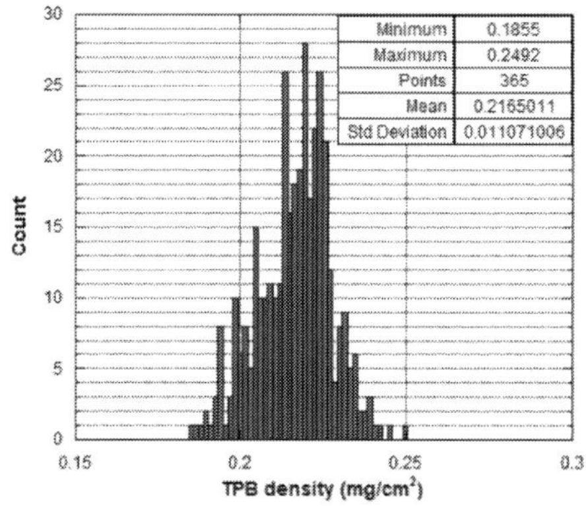

Figure 7. Resulting distribution of the coating densities of the mylar samples. Each sample is related to a PMT evaporation run of the series production. In addition, 5 more PMTs were coated as spare.

6. Conclusions

The obtained evaporation technique was used to coat with the TPB all the 360 PMTs constituting the new ICARUS T600 light collection system. In every PMT evaporation run, a mylar sample was fixed inside the vacuum chamber in a dedicated support in order to check the coating density as in the preliminary phase. The resulting distribution of the coating densities related to all the PMTs of the series production is shown in Fig. 7.
The measured Quantum Efficiency response all over the PMT sensitive window is rather flat. This demonstrates the good reliability of the adopted evaporation technique, which allows to obtain the best uniformity in terms of coating thickness and light yield conversion.

References

[1] A Aguilar *et al* 2001 *Phys. Rev.* D**64** 112007
[2] S Gariazzo *et al* 2017 *J. High Energ. Phys.* 2017:135
[3] R Acciarri *et al* 2015 *arXiv*:1503.01520
[4] M Antonello *et al* 2015 *JINST* **10** P12004
[5] M Antonello *et al* 2013 *Adv. High Energy Phys.* vol. 2013
[6] S Kubota *et al* 1979 *Phys. Rev.* B**20**, 3486
[7] A Falcone 2016 *Ph.D. Thesis*

Demagnetization induced peak effect for a FeSeTe granular superconductor in ac magnetic field

D Mancusi[1], A Galluzzi[1,2], S Pace[1,2] and M Polichetti[1,2]

[1]Department of Physics "E.R. Caianiello", University of Salerno, via Giovanni Paolo II, 132, Fisciano (SALERNO), I-84084, Italy
[2]CNR-SPIN Salerno, via Giovanni Paolo II, 132, Fisciano (SALERNO), I-84084, Italy
E-mail: agalluzzi@unisa.it

Abstract. The possible evidence of a peak effect in the temperature dependent magnetization and in the field dependent critical current density has been found for a FeSeTe superconducting granular sample in presence of an ac field. The existence of the peak effect has been confirmed by the numerical calculation of the inter- and intragranular magnetizations in presence of a demagnetization field with components at the third harmonic frequency generated by the sample magnetization. This suggests that the peak effect could be related to the flux dynamics inside the sample through the demagnetization field third harmonic.

1. Introduction

The critical current density is one of the most useful parameters to be applied to type II superconductors, which is typically estimated by performing the magnetic hysteresis loop in presence of a dc magnetic field at different temperatures [1,2], and within the Bean's critical state model [3]. This allows constructing both the temperature and field dependence of the critical current density [3–5]. The peak effect is one of the most interesting phenomena, which can be observed in both high-T_c and Fe-based superconductors, consisting in the anomalous increase of the critical current density with increasing the applied field due to the pinning changes [6–11]. Although it is typically detected from dc magnetic measurements, the ac magnetic complex susceptibility technique may be also useful since it allows exploring the magnetic response of the samples at short time scales [12–18]. In fact, several interpretations of the peak effect give to it a dynamical origin in the critical state of superconductors [8,10,13,19–23], and the critical state and its characteristic frequency dependent flux dynamical regimes are accessible by performing pulsing field measurements. In presence of an ac field, the critical current density increases before dropping to zero at the superconducting transition, which produces a sudden decrease in the in-phase magnetization $M_1'(T)$ (or in the corresponding susceptibility real part $\chi_1'(T)$). Moreover, depending on the pinning strength, the sample geometry and the applied field amplitude, a second dissipation peak can occur in the out-of-phase magnetization $M_1''(T)$ (or in the corresponding susceptibility real part $\chi_1''(T)$) [24–29].

In the present work, we report the analysis of the ac magnetization measurements as function of the temperature and of the field dependent critical current density of a FeSeTe superconducting granular sample. From the experimental magnetization the first harmonic components of the inter- and intragranular magnetizations have been numerically calculated by using a model proposed in our previous work [30,31] and taking into account the magnetic interaction in the intergranular regions by means of the demagnetization fields at both the fundamental and third harmonic frequencies. In particular, we will relate these fields to the existence of a peak effect in the critical current density extracted from the experimental magnetic curves.

2. Experimental results

The temperature T dependent fundamental and higher harmonics of the time t dependent magnetization M of a $FeSe_{0.5}Te_{0.5}$ granular sample, of the same batch obtained starting from its basic compound FeSe [32,33], have been acquired in absence of dc magnetic fields and in presence of an ac field $\mu_0 H_{ac} \sin(2\pi\nu t)$ (ν-frequency) by means of the ac susceptibility technique. The first five nonzero in-phase and out-of-phase odd components $M_n^{'}$ and $M_n^{''}$ with $n = 1,3,\ldots,9$ were measured as function of the temperature, while zero even component were found as expected in absence of dc field [12,13]. The magnetic field amplitude and frequency were taken as $\mu_0 H_{ac} = 0.5, 1, 2, 4, 8, 12$ mT and $\nu = 107, 1077, 5385, 9693$ Hz. The field was applied perpendicular to the longest dimension of the sample which consisted of a rectangular prism.

The curves of the first harmonic components $M_1^{'}(T)$, $M_1^{''}(T)$ and of the third harmonic components $M_3^{'}(T)$, $M_3^{''}(T)$ of the sample magnetization are reported in the Fig. 1a-1b and Fig. 1c-1d, respectively, at different amplitudes $\mu_0 H_{ac}$ and at fixed frequency ν of the applied ac field. The curves of the higher harmonics components are not reported since they are at least one order of magnitude lower than the first and third ones, and then can be neglected in the aim of the investigation of the ac response of the sample. The curves of $M_1^{''}(T)$ exhibit a large peak and a small peak at intermediate and high temperatures, respectively. At a first sight, one could ascribe these two magnetic contributions to the inter- and intragranular ac responses, respectively [12,13,34]. However, as shown in the inset of the Fig. 1b, both peaks shift to lower temperatures with the increase of the field, with almost the same trend, while the intergranular contribution is expected to be faster than the intragranular one due to the different strengths of the superconductivity in the corresponding volume fractions [12,13,34,35]. However, after a more precise analysis of the curves and of their temperature derivatives (not reported here), we have isolated an additional very slight peak at low temperatures. Since this peak shifts very fast when increasing the ac field amplitude, as shown in the inset of the Fig. 1b, we argue that it could be ascribed to the intergranular contribution, indicating a bad quality of the intergranular links of the sample which produces a high coherence of the superconducting transition inside the sample [12]. Within such interpretation the smaller peak near the transition temperature T_c should be ascribed to the intragranular response and the second larger one at intermediate temperatures T_p could be an additional effect.

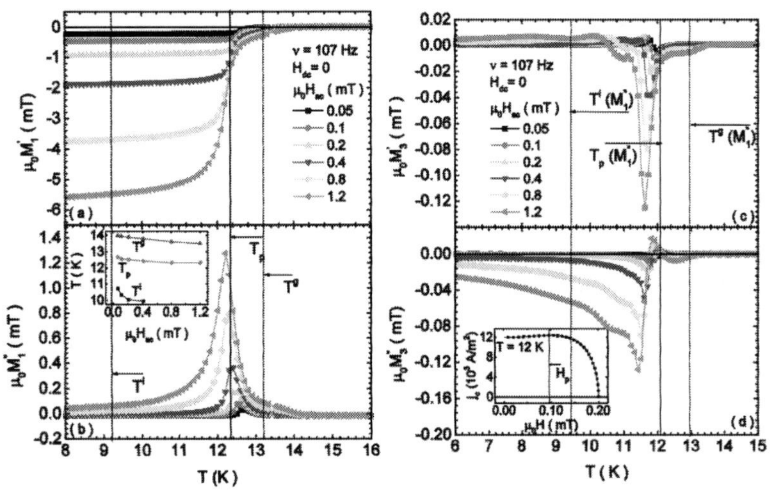

Figure 1. (Left) Plots of the temperature dependent magnetization first harmonic (a) in-phase and (b) out-of-phase components measured on the $FeSe_{0.5}Te_{0.5}$ sample, at different amplitudes $\mu_0 H_{ac}$ and at fixed frequency ν of the applied ac field H. The vertical lines indicate (dotted) the intergranular T^i and (dashed) the intragranular T^g dissipation peaks temperatures, and (solid) the possible peak effect T_p temperatures, and the inset is the plot of the field dependence of the peaks positions. (Right) Plot of the corresponding third harmonic curves with the inset showing the field dependent critical current density of the sample calculated from the $M(H)$ loops constructed by taking the magnetization curves of the (a,b,c,d) panels and the applied field $\mu_0 H_{ac} \sin(2\pi\nu t)$[3,4] ($H_p$ field of the peak effect).

In order to further analyze the experimental magnetization curves, the field dependent critical current density has been extracted from the curves of the magnetic hysteresis loops of the system constructed by taking the experimental time dependent magnetization signal $M(t,T) = \sum_{n=1,3} M_n(T,t)$ with $M_n(T,t) = M_n^{'}(T)\sin(n\omega t) + M_n^{''}(T)\cos(n\omega t)$ (with $\omega = 2\pi\nu$), as function of the applied field $\mu_0 H_{ac}\sin(\omega t)$, at different temperatures. From these loops, within the Bean's critical state model [3,4], the critical current density j_c as function of the applied field during an ac field cycle has been calculated as the difference between the demagnetization and magnetization branches of the loop normalized to a factor related to the geometry of the sample's surface perpendicular to the applied field. The curve of $j_c(H)$ at intermediate amplitude and frequency values of the applied ac field and at the corresponding temperature T_p of the possible peak effect in the Fig. 1b, is reported in the inset of the Fig. 1d. This curve clearly exhibits an increase of the current density with the increase of the field at intermediate values of the field, as expected in presence of a peak effect in the magnetic response of the sample. In order to identify the inter-and intragranular nature of the peak effect, the inter- (i) and intragranular (g) ac magnetization curves $M_1^{i,g\,'}(T)$, $M_1^{i,g\,''}(T)$ and ac susceptibility curves $\chi_1^{i,g\,'}(T)$, $\chi_1^{i,g\,''}(T)$ will be determined and analyzed in the following section.

3. Analysis and discussions of the fundamental magnetization

In order to investigate the actual origin of the peak effect appearing in the curves of the critical current density of the sample reported in the inset of the Fig. 1d, the fundamental magnetization curves $M_1^{'}(T)$ and $M_1^{''}(T)$ have been analyzed in the framework of a model for the description of the ac response of a granular system [30]. In fact, the inter- and intragranular magnetization and susceptibility curves have been calculated from the whole sample's experimental curves, by numerically solving the equation (μ_0 vacuum magnetic permeability)

$$\mu_0 M_n(t,T) \approx \frac{(1-f^g+D^i)\mu_0 M_n^i(t,T)+(f^g+D^g)\mu_0 M_n^g(t,T)}{1-D},\ n = 1,3. \quad (1)$$

Here f^g is the granular volume fraction of the sample, D is the demagnetization factor for the whole sample geometry, while D^i and D^g are two effective inter- and intragranular demagnetization factors. Moreover, $M_n^i(T)$ and $M_n^g(T)$ are the inter- and intragranular magnetizations Fourier components

$$\mu_0 M_n^{i,g}(t,T) = \mathrm{Im}\left\{\sum_{n=1,3} \frac{\chi_n^{i,g}}{\pi}\left[B^{i,g}(t,T)\right]^n\right\}, \quad (2)$$

where $\chi_n^{i,g}$ are the susceptibilities harmonics, and the demagnetization effects due to the geometry of the whole sample and of the grains are taken into account within two effective magnetic fields $B^i(t,T)$ and $B^g(t,T)$ acting at the surfaces of the sample and of the grains, respectively. These are given by

$$B^i(t,T) = \mu_0 H_{AC} e^{i\omega t} - D\{\mu_0[M_1^{'}(T)e^{i\omega t} + iM_1^{''}(T)e^{-i\omega t}] + [M_3^{'}(T)e^{3i\omega t} + iM_3^{''}(T)e^{-3i\omega t}]\}, \quad (3a)$$

$$B^g(t,T) = \mu_0 H_{AC} e^{i\omega t} + (1-D)\{\mu_0[M_1^{'}(T)e^{i\omega t} + iM_1^{''}(T)e^{-i\omega t}] + [M_3^{'}(T)e^{3i\omega t} + iM_3^{''}(T)e^{-3i\omega t}]\}. \quad (3b)$$

It is worth noting that the first and the third harmonics of the sample magnetization $M_1(t,T)$ and $M_3(t,T)$ generate two demagnetization fields contributions at the fundamental frequency and two contributions at the third harmonic frequency. These fields at frequency ω and 3ω produce contributions in the fundamental magnetizations, named $M_{11}^{i,g}(t,T)$ and $M_{31}^{i,g}(t,T)$, respectively, for the inter- and intragranular volume fractions. The factor D can be taken from the literature for the considered sample geometry.

The inter- and the intragranular magnetization curves, in presence of the demagnetization field harmonics components at the fundamental frequency $-D[M_1^{'}(T)e^{i\omega t} + iM_1^{''}(T)e^{-i\omega t}]$ and $(1-D)[M_1^{'}(T)e^{i\omega t} + iM_1^{''}(T)e^{-i\omega t}]$ only and in presence of both the fundamental components and the components at the third harmonic frequency $-D[M_3^{'}(T)e^{i\omega t} + iM_3^{''}(T)e^{-i\omega t}]$ and $(1-D)[M_3^{'}(T)e^{i3\omega t} +$

$iM_3''(T)e^{-i3\omega t}$], have been numerically calculated from the self-consistent equations system of the Eq. (1), together with the Eq. (2) and (3a)-(3b). The corresponding susceptibilities $\chi_1^{i'}(T)$, $\chi_1^{i''}$ and $\chi_1^{g'}(T)$, $\chi_1^{g''}$, have also been obtained within the calculation.

The curves of $M_{11}^{i,g'}(T)$, $M_{11}^{i,g''}(T)$ and $M_{31}^{i,g'}(T)$, $M_{31}^{i,g''}(T)$ are shown in the Fig. 2a-2b and 2c-2d, respectively, together with the corresponding susceptibility curves. In these Fig., the in-phase magnetization curves $M_{11}^{i}{}'(T)$ and $M_{11}^{g}{}'(T)$ exhibit the typical behavior expected in absence of a peak effect in ac field, that is a negative peak due to the temperature dependence of the amplitudes $B^g(T)$ of the effective field acting at the surfaces of the grains which is shielded by the intergranular currents at lower temperatures. Moreover, two peaks appear in the corresponding out-of-phase curves $M_{11}^{i}{}''(T)$ and $M_{11}^{g}{}''$, respectively, at the temperatures T^i and T^g individuated in the Fig. 1a and 1b.

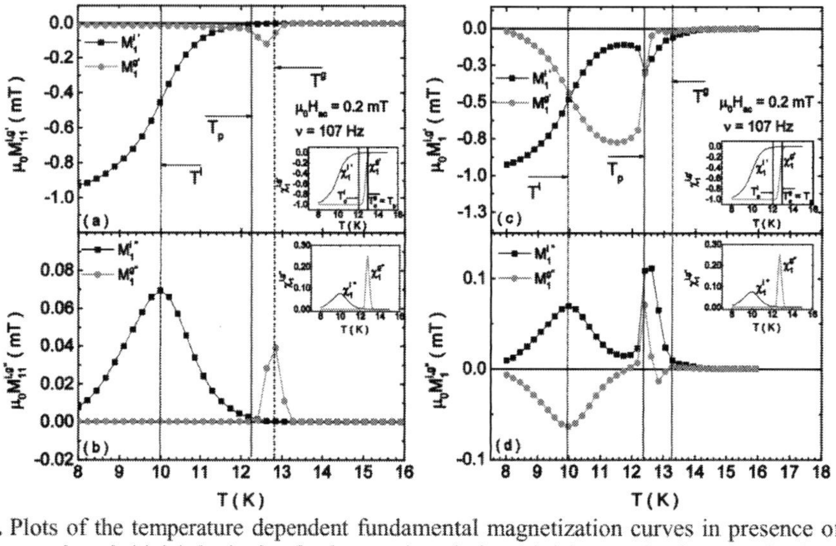

Figure 2. Plots of the temperature dependent fundamental magnetization curves in presence of (left) the fundamental and (right) both the fundamental and the third harmonic demagnetization fields calculated within the model described in the previous section and in Ref.[30]. The (a,c) and the (b,d) panels are the in-phase and the out-of-phase components, respectively.

On the other hand, no effects are visible at the peak effect temperature T_p, showed in the curves of the Fig. 1a-1b and of the inset of the Fig. 1d, containing all the magnetization amounts acquired by the measuring system. Since these curves exhibit the magnetic response of the sample to the applied field and to the demagnetization field at both the fundamental and the higher frequencies, this suggests that the peak effect occurring in the magnetic response could be not visible in the curves of the fundamental magnetization components of the Fig. 2a-2b due to the fact that only the fundamental demagnetization field has been considered in their calculation. This kind of interpretation is confirmed from the inter- and the intragranular magnetization curves of the Fig. 2c-2d obtained when considering the demagnetization field third harmonics. Here it is clearly visible the existence of a drop in both the inter- and intragranular in-phase curves $M_1^{i'}(T)$ and $M_1^{g'}(T)$, near the corresponding transition temperatures T_c^i and T_c^g, respectively. At these temperatures, a second peak is observed in the inter- and intragranular out-of-phase curves $M_1^{i''}(T)$ and $M_1^{g''}(T)$. Both these features of the in-phase and out-of-phase curves could be considered as the evidence, in presence of an ac field, of a peak effect in both the inter- and intragranular volume fractions. In this kind of interpretation, the intergranular peak effect could be covered in the curves of the whole sample magnetization of the Fig. 1a-1b, while the intragranular peak effect remains visible in the out-of-phase magnetization curves.

4. Conclusions

The fundamental harmonics components of the magnetizations of the inter- and intragranular volume fractions in a superconducting granular sample have been numerically extracted from the experimental whole sample magnetization as function of the temperature. The experimental curves indicate the existence of a peak effect in the critical current density of the sample, and the features of this phenomenon have been found in the separated inter- and intragranular magnetization components only when considering the existence of demagnetization fields components at the third harmonic frequency in addition to the fundamental harmonics components. These results suggest the third harmonic demagnetization field at the origin of the peak effect observed in the critical current density of the superconducting granular sample. This could be crucial since within the used model the demagnetization induced peak effect depends on the granularity of the system and then it could be controlled in the applications by means of the sample granular structure.

References

[1] Galluzzi A, Polichetti M, Buchkov K, Nazarova E, Mancusi D and Pace S 2015 *Supercond. Sci. Technol.* **28** 115005
[2] Galluzzi A, Polichetti M, Buchkov K, Nazarova E, Mancusi D and Pace S 2017 *Supercond. Sci. Technol.* **30**,025013
[3] Bean C P 1964 *Rev. Mod. Phys.* **36** 31–9
[4] Bean C P 1962 *Phys. Rev. Lett.* **8** 250–3
[5] Leo A, Grimaldi G, Guarino A, Avitabile F, Nigro A, Galluzzi A, Mancusi D, Polichetti M, Pace S, Buchkov K, Nazarova E, Kawale S, Bellingeri E and Ferdeghini C 2015 *Supercond. Sci. Technol.* **28** 125001
[6] Cohen L F, Laverty J R, Perkins G K, Caplin A D and Assmus W 1993 *Cryogenics (Guildf).* **33** 352–6
[7] Jirsa M, Půst L, Dlouhý D and Koblischka M R 1997 *Phys. Rev. B* **55** 3276–84
[8] Abulafia Y, Shaulov A, Wolfus Y, Prozorov R, Burlachkov L, Yeshurun Y, Majer D, Zeldov E, Wühl H, Geshkenbein V B and Vinokur V M 1996 *Phys. Rev. Lett.* **77** 1596–9
[9] Nishizaki T, Naito T and Kobayashi N 1998 *Phys. Rev. B* **58** 11169–72
[10] Li D P, Lin P-J, Rosenstein B, Shapiro B Y and Shapiro I 2006 *Phys. Rev. B* **74** 174518
[11] Safar H, Gammel P L, Bishop D J, Mitzi D B and Kapitulnik A 1992 *Phys. Rev. Lett.* **68** 2672–5
[12] Gömöry F 1997 *Supercond. Sci. Technol.* **10** 523–42
[13] Hein R, Francavilla T and Liebenberg D 1991 *Magnetic Susceptibility of Superconductors and Other Spin Systems* ed R A Hein, T L Francavilla and D H Liebenberg (Boston, MA: Springer US)
[14] Polichetti M, Adesso M G and Pace S 2004 *Physica A: Statistical Mechanics and its Applications* **339** 119–24
[15] Zola D, Gomory F, Polichetti M, Strycek F, Seiler E, Husek I, Kovac P and Pace S 2004 *Supercond. Sci. Technol.* **17** 501–11
[16] Ishida T and Goldfarb R B 1990 *Phys. Rev. B* **41** 8937–48
[17] Van Duyneveldt A J 1982 *J. Appl. Phys.* **53** 8006–11
[18] Senatore C, Polichetti M, Clayton N, Flükiger R and Pace S 2004 *Phys. C Supercond. its Appl.* **401** 182–6
[19] Miu D, Noji T, Adachi T, Koike Y and Miu L 2012 *Supercond. Sci. Technol.* **25** 115009
[20] Larkin A I, Marchetti M C and Vinokur V M 1995 *Phys. Rev. Lett.* **75** 2992–5
[21] Bonura M, Giannini E, Viennois R and Senatore C 2012 *Phys. Rev. B* **85** 134532
[22] Ghosh K, Ramakrishnan S, Grover A K, Menon G I and Chandra G 1996 *Phys. Rev. Lett.* **1** 1–4
[23] Kwok W K, Fendrich J A, Van Der Beek C J and Crabtree G W 1994 *Phys. Rev. Lett.* **73**

2614–7
[24] Bossen O, Schilling A and Toyota N 2013 *Phys. C Supercond.* **492** 133–7
[25] Satyajit 2011 Nonlinear response of the static and dynamic phases of the vortex matter *Supercond. - Theory Appl.* **4** 1-30
[26] Mohan S, Sinha J, Banerjee S S and Myasoedov Y 2007 *Phys. Rev. Lett.* **98** 1–4
[27] Patil N G, Banerjee S S, Saha S, Ghosh K, Ramakrishnan S, Bhattacharya S, Grover A K, Rao T V C, Ravikumar G, Mishra P K, Sahni V C, Yamamoto E, Haga Y, Hedo M, Inada Y and Onuki Y 1997 *Phys. C* **282** 2043–4
[28] Inoue K, Sakai N, Waki K and Murakami M 2001 *Phys. C Supercond.* **357–360** 531–3
[29] Ge J, Gutierrez J, Li J, Yuan J, Wang H B, Yamaura K, Takayama-Muromachi E and Moshchalkov V 2013 *Phys. Rev. B - Condens. Matter Mater. Phys.* **88** 1–7
[30] Mancusi D, Galluzzi A, Pace S and Polichetti M 2017 *J. Phys. Condens. Matter* **29** 425701
[31] Mancusi D, Polichetti M, Cimberle M R and Pace S 2015 *Supercond. Sci. Technol.* **28** 095017
[32] Hsu F-C, Luo J-Y, Yeh K-W, Chen T-K, Huang T-W, Wu P M, Lee Y-C, Huang Y-L, Chu Y-Y, Yan D-C and Wu M-K 2008 *Proc. Natl. Acad. Sci.* **105** 14262–4
[33] Fiamozzi Zignani C, Corato V, Leo A, De Marzi G, Mancini A, Takano Y, Yamashita A, Polichetti M, Galluzzi A, Rufoloni A, Grimaldi G and Pace S 2016 *IEEE Trans. Appl. Supercond.* **26** 1–5
[34] Aste T, Di Matteo T and Tordesillas A 2007 Granular and Complex Materials **8** (World Scientific Publishing Co. Pte. Ltd.)
[35] Polichetti M, Zola D, Luo J-L, Chen G-F, Li Z, Wang N-L, Noce C and Pace S 2012 *Supercond. Sci. Technol.* **25** 025010

IOP Conf. Series: Journal of Physics: Conf. Series **956** (2017) 012018 doi:10.1088/1742-6596/956/1/012018

Coarsening and percolation in the Ising Model with quenched disorder

F Insalata[1,3], F Corberi[2], L F Cugliandolo[3] and M Picco[3]

[1] Dipartimento di Fisica "E. R. Caianiello", Università di Salerno, via Giovanni Paolo II 132, 84084 Fisciano (SA), Italy.

[2] INFN, Gruppo Collegato di Salerno, and CNISM, Unità di Salerno, Università di Salerno, via Giovanni Paolo II 132, 84084 Fisciano (SA), Italy.

[3] Sorbonne Universités, Université Pierre et Marie Curie - Paris 6, Laboratoire de Physique Théorique et Hautes Energies, 4, Place Jussieu, Tour 13, 5ème étage, 75252 Paris Cedex 05, France.

E-mail: `ferdinsa@live.it`

Abstract. Through large-scale numerical simulations, we study the phase ordering kinetics of the $2d$ Ising Model after a zero-temperature quench from a high-temperature homogeneous initial condition. Analysing the behaviour of two important quantities – the winding angle and the pair-connectedness – we reveal the presence of a percolating structure in the pattern of domains. We focus on the pure case and on the random field and random bond Ising Model.

1. Introduction

Phase ordering kinetics, the ordering of a system via domain growth after a quench from the homogeneous phase into one with broken symmetry, has attracted great interest in the last 50 years [1].

A simple example is a ferromagnet, instantaneously cooled (quenched) from above to below the critical point. The initial equilibrium state becomes unstable after the quench and evolves towards one of the two possible symmetry-related ordered configurations with opposite magnetizations. Relaxation toward the new equilibrium state occurs slowly (i.e. not exponentially fast) by the formation and growth (coarsening) of domains of the two equilibrium phases (i.e. group of aligned spins). This domain growth is driven by superficial tension, i.e. the interfaces tend to become flatter due to energetic reasons. Over time, the smallest domains disappear so that the typical size $R(t)$ of the remaining ones increases.

One theoretical approach to this class of problem is the kinetic Ising Model (IM), originally introduced by Glauber [2]. One starts with a Ising system in which each spins is randomly oriented, a situation that can be described as equilibrium at infinite temperature, $T \to \infty$. Then, the evolution of the system is treated as a Markov chain, with appropriate transition probabilities. The dynamical evolution can be simulated through standard Monte Carlo methods [3]. In the thermodynamic limit, the coarsening process goes on indefinitely, with none of the two equilibrium phases (up and down spins in this case) prevailing at any given finite time. This means that the magnetic system does not develop a magnetization, or equivalently that on average up spins are in the same number as down spins.

It was recently shown [4, 5, 6] that, after a transient from the quench, some of the morphological features of the coarsening domains in an ordinary, homogeneous IM are the same as those of a geometrical lattice at the percolation threshold p_c of random site percolation [7]. In such condition, a spanning (percolating) cluster is present. In the coarsening IM, we observe a percolating domain of either up or down spins. This fact may seem paradoxical, since percolation is a non-interacting problem, whereas coarsening is driven by spin-spin interactions. Moreover, the absence of magnetization in the coarsening system means that the percolative structures are present when 1/2 of spins are up (down). For a $2d$ square lattice, the same geometry we will consider for the IM, the critical percolation threshold is $p_c \simeq 0.59$, when occupying sites randomly. This means that we observe percolation in the corsening IM when the occupation probability of up (down) spins is $1/2 < p_c$. This is not a contradiction, however, since after the transient t_p, the spin configuration is no longer random, having been shaped by the spin-spin interactions. It is, indeed, only after the coarsening dynamics has gone on for a transient that percolative features appear, while they are absent right after the quench, a sign that they are due to interactions. This can be phrased by saying that the spin-spin interactions lower the percolation threshold from $p_c \simeq 0.59$ to 1/2. Moroever, although random uncorrelated percolation and correlated coarsening coexist on the same lattice, they live on separate length scales : the percolative properties are not present, at any given time, on length scales that are smaller than $R(t)$, the typical size of the domains, i.e. the correlated regions.

In a recent work [8], we showed that this phenomenology is also found in two types of inhomogeneous IM, where structural (quenched) disorder is present. In particular, we focussed on the random field Ising Model (RFIM) and random bond Ising Model (RBIM).

In this paper, after a brief review of the models considered and of the Monte Carlo algorithm implemented (Sec. 2), we define the relevant observables in Sec. 3 and present the basic results which highlight the occurrence of the percolation transition in Sec. 4.

2. Models and algorithm

We consider a $2d$ square lattice of linear size L, and the Hamiltonian of our models is

$$\mathcal{H}(\{S_i\}) = -\sum_{\langle ij \rangle} J_{ij} S_i S_j + \sum_i H_i S_i, \tag{1}$$

where S_i are the Ising variables and $\langle ij \rangle$ are two nearest-neighboring sites. The different models considered in our study are determined by the properties of the coupling constant J_{ij} and of the external field H_i, as follows.

Homogeneous IM: In this case, $H_i \equiv 0$ and $J_{ij} \equiv J_0$, i.e. we recover the ordinary IM. We also refer to it as the clean or pure model.

RFIM: The couplings are constant $J_{ij} \equiv J_0$, while the external fields H_i are drawn from a symmetric bimodal distribution and uncorrelated in space, i.e. $H_i = \pm h$, each with probability one half. We refer to this model as the RFIM.

RBIM: No external fied, $H_i \equiv 0$, while the couplings are $J_{ij} = J_0 + \delta_{ij}$, where δ_{ij} are independent random numbers drawn from a uniform distribution in $[-\delta, +\delta]$. We keep $\delta < J_0$ in order to avoid frustration effects. We refer to this model as the RBIM.

In the case of the RBIM, coarsening via domain growth still takes place in the absence of frustration, which is the case we have considered. In the case of the RFIM, the Imry-Ma argument [9] tells us that ordering takes place up to a length

$$\ell_{IM} \sim \left(\frac{h}{J_0}\right)^{\frac{2}{d-2}}. \tag{2}$$

As we will see below, we focus on the case in which $\ell_{IM} \to \infty$, for which the system orders in any dimension.

As mentioned above, we can treat the dynamical evolution of the system as a Markov chain and simulate it through the Metropolis algorithm. Specifically, our algorithm flips single spins with Glauber transition rate

$$w(S_i \to -S_i) = \frac{1}{2}\left[1 - S_i \tanh\left(\frac{H_i^W + H_i}{T}\right)\right], \tag{3}$$

where the Weiss Field H_i^W, is given by

$$H_i^W = \sum_{j \in nn(i)} J_{ij} S_j, \tag{4}$$

with j running over the nearest-neighbors of the spin S_i.

For our simulations, the protocol consists in preparing a system of uncorrelated spins with no magnetization – equilibrium at $T \to \infty$ – and letting it evolve through single-spin flip according to Eq. (3). Performing the quench to T_f amounts to inserting the value of T_f into Eq. (3). We have let $T_f \to 0$ keeping the ratios $\epsilon = \frac{\delta}{T_f}$ (for RBIM) and $\epsilon = \frac{h}{T_f}$ (for RFIM) finite.

First, let us notice that this limit entails $\ell_{IM} \to \infty$, as mentioned above. Moreover, this simplifies the form of the transition rates, which depend only on the parameter ϵ. Not only we obtain a theory that depends on a single parameter, but simulations are sped up and thermal noise is reduced. More details on this algorithm and on its reliability can be found in [10, 11].

3. Key observables

The first quantity we consider, an essential one for the study of the phase ordering kinetics of a system, is the typical domain size $R(t)$. This is also referred to as *growing length*, since it grows while the system orders through domain growth. We have computed it as the inverse of the density of defects, i.e. dividing the number of couple of unaligned spins by L^2. As discussed in [12], this is a standard method to estimate $R(t)$. We have then performed, as for all other quantities, an ensemble average, typically on $10^5 - 10^6$ realizations. $R(t)$ grows as $t^{1/2}$ in the pure system and slower and slower for increasing values of the parameter ϵ controlling the strength of disorder. More details can be found in [13].

In order to probe the emergence of percolation in the coarsening process, we used two observables : the average squared winding angle and the pair-connectedness function. They are naturally defined on a geometrical lattice in which a fraction p of the sites is occupied. They can be measured on an Ising lattice by identifying occupied sites as, say, up spins and empty sites as down spins, or vice-versa. A domain can then be considered as a cluster of up (or down) spins and the above quantities can be computed with respect to the domains of the system.

The (average squared) winding angle is defined considering the external contours – hulls – of a cluster of occupied sites. One considers two points P and Q at a distance x along the hull, and the tangent to the hull in these two points. The angle between the two tangents, measured counterclockwise in radiants units, is the winding angle $\theta(P, Q)$. By fixing x, one can calculate the average winding angle for two points separated by this distance, $\langle\theta(x)\rangle$. It can be exactly proven, through methods of conformal field theory, that on a lattice at the percolation threshold for the hull of any cluster we have [14, 15]

$$\langle\theta^2(r)\rangle = a + \frac{4k}{8+k}\ln r, \tag{5}$$

where a is a constant and $k = 6$. This equation holds when $r \gg r_0$, r_0 being the lattice spacing, i.e. when the microscopic structure of the lattice can be ignored. Therefore, we will report numerical data only for the largest domain of the system, so to have the largest possible range in which $r \gg r_0$.

The pair-connectedness function is defined in percolation theory as the probability that two occupied sites at distance r belong to the same cluster. From $2d$ percolation theory, we know that at $p = p_c$ [16, 17]

$$C_{perc}(r, r_0) \sim \left(\frac{r}{r_0}\right)^{-2\Delta} \tag{6}$$

where r_0 is the unit of length, usually the lattice spacing. Eq. (6) holds for $r \gg r_0$, with the critical exponent $\Delta = 5/48$. In the coarsening system we measure the pair-connectedness at any given time, identifying all the domains of up and down spins and then computing

$$C(r,t) = \frac{1}{4L^2} \sum_i \sum_{i_r} \langle \delta_{S_i, S_{i_r}} \rangle, \tag{7}$$

where $\delta_{S_i,S_j} = 1$ if the two spins belong to the same domain – namely they are aligned and there is a path of aligned spins connecting them – and $\delta_{S_i,S_j} = 0$ otherwise.

4. Numerical results

We now show how measuring the winding angle and the pair-connectedness on the domains of the coarsening IM highlights that the morphological properties of the domain patchwork are those of critical percolation.

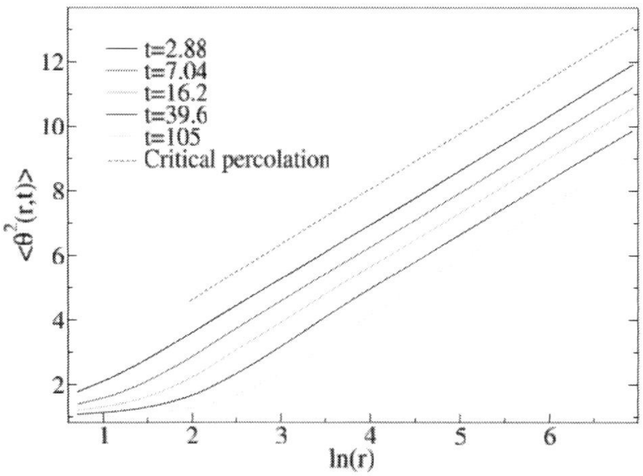

Figure 1. Average squared winding angle for the domains of the largest cluster of the pure $2d$ IM, plotted against $\ln r$, the distance r along the hull. Results are relative to a lattice with $L = 640$. In the key we report the measurement times. The dashed line is Eq. (5) with $k = 6$, the value of critical percolation.

In Fig. 1 we report results for the clean model of size $L = 640$, for various measurement times. The dashed blue line is the analytical curve, namely Eq. (5) with $k = 6$.

Apart from an upward vertical shift of the curves for increasing measurement times – whose interpretation is discussed in [8] – for each curve we observe two behaviors : an initial part (short distances) in which the analytical expression is not followed, and a large-distance regime where numerical data are very well described by Eq. (5). Indeed, we fitted the slope of the purple curve corresponding to $R(t) = 2.68$ and $t = 3.13$ in the range $4 < \ln(x) < 7$, and we obtained $k = 5.94$, in excellent agreement with the value $k = 6$ for critical percolation. This signals that over sufficiently large distances the largest cluster has the morphological properties of percolation. Interestingly, we see that the crossover between the two behaviors takes place at a distance l_t which increases with the measurement time.

We find a similar behaviour for pair-connectedness, shown for the pure IM in Fig 2. We performed the measurement of $g(r)$ at different times on a system of size $L = 512$. For each curve except $t = 2$, after an initial distance which increases with time, curves for coarsening have the same slope as that for percolation, the black line with slope $-2\Delta = -10/48$ (at large distances the curves bends upward as due to finite size effects). Notice that this is not true for very small times, when the system is still very close to the initial condition. The vertical shift of the curves for varying measurement times is interpreted in [8] in the same way as for the winding angle.

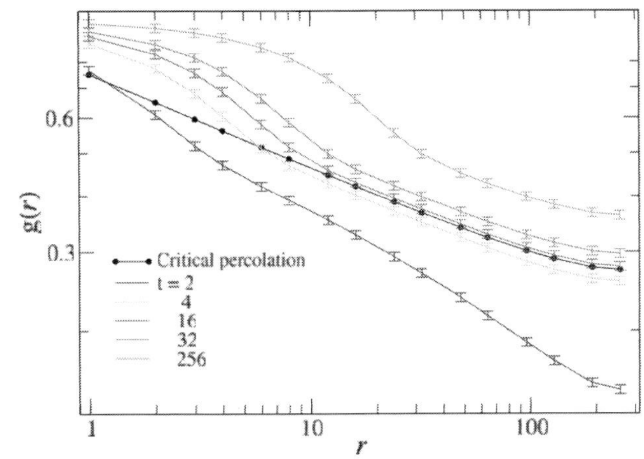

Figure 2. Pair connectivity measured on a lattice at percolation threshold (black line) and for a coarsening homogeneous $2d$ IM (colored lines) is plotted against the distance r. Measurement times are given in the key. The lattice size is $L = 512$ in both cases.

In Figs. 1 and 2 we noticed a length, growing with time, after which the domains acquire the morphological properties of critical percolation. It is natural to identify this length with the growing length $R(t)$, the typical size of the correlated regions (domains). Indeed, as mentioned in Sec. 1, percolative properties can only be present a length scale over which the system is still uncorrelated. This interpretation can be formalised and quantitatively verified through scaling arguments [8].

Most importantly, we find analogous results for all the values of disorder strength considered, for both RFIM and RBIM. We only report data for the RFIM with $\epsilon = 1$ in Figs. 3 and 4

We regard this as compelling evidence that the clusters arising in the coarsening process *are* those of critical percolation, for both pure and disordered systems. We can therefore conclude that percolation effects are quite robust, since they are not destroyed by the presence of the considered quenched disorder.

5. Conclusions

In this work we have investigated the relevance of percolative effects on the coarsening process of the $2d$ IM quenched from a completely disordered initial state to zero final temperature. We focused on the clean case and on two forms of quenched randomness, random bonds and random fields. We have used two quantities – winding angle and pair-connectedness – which represent efficient tools to detect percolation effects.

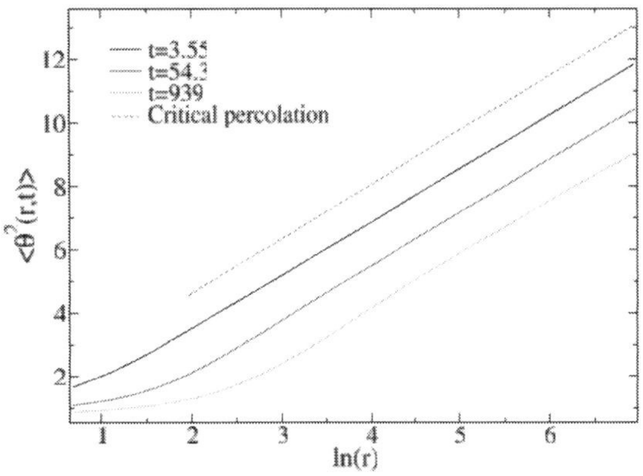

Figure 3. Same as Fig. 1, but for RFIM with $\epsilon_{RF} = 1$.

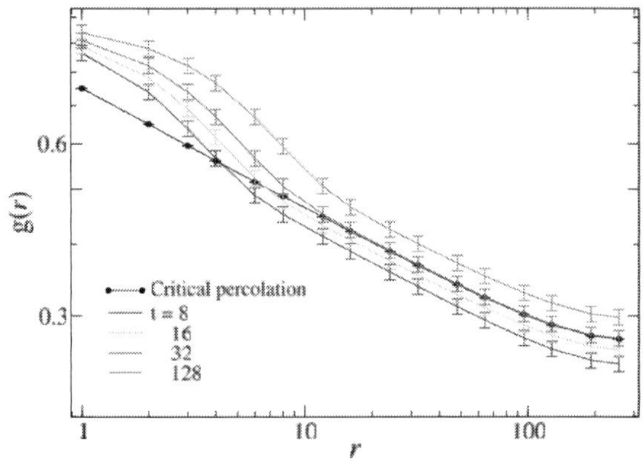

Figure 4. Same as Fig. 2 but for a RFIM with $\epsilon_{RF} = 1$.

In the light of our results, we can conclude that the influence of percolation on the coarsening dynamics of a quenched ferromagnet – previously highlighted for homogeneous models – also extends to systems where quenched disorder is present, systems that are much less understood. These results represent a first necessary step toward the generalisation of one of the few analytical theories of coarsening [18] – currently available for pure systems – that hinges on the percolative properties of the domains of the ordering system.

Finally, a natural extension of this work concerns randomly diluted models, a class of systems whose phase ordering kinetics presents subtleties and differences with respect to random field and random bond models [19].

Acknowledgments

F. I. and F. C. thanks LPTHE Jussieu for hospitalty. L. F. C. is a member of Institut Universitaire de France, and she thanks the KITP University of California at Santa Barbara for hospitality. This research was supported in part by the National Science Foundation under Grant No. PHY11-25915.

References

[1] Bray A J 1994 *Adv. Phys.* **43** 357
[2] Glauber R J 1963 *J. Math. Phys.* **4** 294
[3] Cugliandolo L F 2015 *Comptes rendus - Physique* **16** 257
[4] Arenzon J J, Bray A J, Cugliandolo L F and Sicilia A 2007 *Phys. Rev. Lett.* **98** 145701
[5] Sicilia A, Arenzon J J , Bray A J, and Cugliandolo L F 2007 *Phys. Rev.* E **76** 061116
[6] Blanchard T, Corberi F, Cugliandolo L F and Picco M 2014 *EPL* **106** 66001
[7] Christensen K and Moloney N R 2005 *Complexity and Criticality* (London: Imperial College Press)
[8] Corberi F, Insalata F, Picco M and Cugliandolo L F 2017 *Phys. Rev.* E **95** 022101
[9] Imry Y and Ma S-K 1975 *Phys. Rev. Lett.* **35** 1399
[10] Burioni R, Cassi D, Corberi F and Vezzani A 2007 *Phys. Rev.* E **75** 011113
[11] Corberi F, Lippiello E and Zannetti M 2001 *Phys. Rev.* E **63** 061506
[12] Puri S and Wadhawan V. 2009 *Kinetics of Phase Transitions* (Boca Raton: CRC Press)
[13] Corberi F 2015 *Comptes rendus - Physique* **16** 332
[14] Duplantier B and Saleur H 1998 *Phys. Rev. Lett.* **60** 2343
[15] Wieland B and Wilson D B 2003 *Phys. Rev.* E **68** 056101
[16] Stauffer D and Aharony A 1994 *Introduction to Percolation Theory* (London: Taylor and Francis).
[17] Saberi A A 2015 *Phys. Rep.* **578** 1
[18] Sicilia A, Arenzon J J, Bray A J and Cugliandolo L F 2008 *EPL* **82** 10001
[19] Corberi F, Lippiello E, Mukherjee A, Puri S and Zannetti M 2013 *Phys. Rev.* E **88** 042129

The role of the substrate in Graphene/Silicon photodiodes

G Luongo[1,2], F Giubileo[2], L Iemmo[1] and A Di Bartolomeo[1]

[1]Physics Department, University of Salerno, 84084 Fisciano, Salerno, Italy.
[2]CNR-SPIN, 84084 Fisciano, Salerno, Italy

E-mail: adibartolomeo@unisa.it

Abstract. The Graphene/Silicon (Gr/Si) junction can function as a Schottky diode with performances strictly related to the quality of the interface. Here, we focus on the substrate geometry and on its effects on Gr/Si junction physics. We fabricate and study the electrical and optical behaviour of two types of devices: one made of a Gr/Si planar junction, the second realized with graphene on an array of Si nanotips. We show that the Gr/Si flat device exhibits a reverse photocurrent higher than the forward current and achieves a photoresponsivity of 2.5 A/W. The high photoresponse is due to the charges photogenerated in Si below a parasitic graphene/SiO_2/Si structure, which are injected into the Gr/Si junction region. The other device with graphene on Si-tips displays a reverse current that grows exponentially with the bias. We explain this behaviour by taking into account the tip geometry of the substrate, which magnifies the electric field and shifts the Fermi level of graphene, thus enabling fine-tuning of the Schottky barrier height. The Gr/Si-tip device achieves a higher photoresponsivity, up to 3 A/W, likely due to photocharge internal multiplication.

1. Introduction

Schottky diodes are formed by a metal/semiconductor junction. They possess a low forward voltage drop and fast switching, which make them suitable for fast digital logic circuit, high sensible photodetectors and easy-to-fabricate solar cells. Their performance and photosensitivity are strictly related to the metal electrode [1]. The metal can reflect part of the light or absorb a certain range of the spectrum, which become unavailable for sensing. Such problem can be overcome with the use of a transparent and conductive two-dimensional material such as graphene, in graphene/semiconductor junctions [2]. Graphene is formed by a single layer of carbon atoms arranged in a honeycomb lattice; it possesses maximum surface to volume ratio, high mobility [3, 4], high thermal [5] and electrical conductivity [6], constant optical absorption over a wide spectral range [2], low contact resistance [7, 8, 9], and easy down-scaling [10]. The conduction and valence bands of graphene touch in 6 points, where they form double-cone structures, which correspond to vanishing energy gap and density of states. Then, graphene/semiconductor junctions offer the opportunity to study the phenomena that occurs at the interface between a gapless 2D material and a semiconducting 3D material. In particular, the Gr/Si junction is one of the simplest device that one can realize. It behaves like a Schottky diode and its current-voltage characteristics can be described by thermionic emission theory. Some peculiarities, such as the growing reverse bias saturation current has been explained by the modulation of the Schottky barrier height caused by the low density of states of graphene [2]. In this work, we mainly study how the geometry of the Si substrate can affect the electrical transport and the photoresponse of Gr/Si junctions, a topic which is attracting growing attention [11].

2. Materials and method

We realized two different types of devices. The first device type, Fig. 1(a), was fabricated by depositing a layer of ~ 245 nm thickness of SiO_2 on a low doped n-type Si wafer ($\sim 4.5 \cdot 10^{14}$ cm^{-3}) by chemical vapor deposition (CVD). We patterned a trench of 10 µm wide by lithography, then we removed the SiO_2 from the trench area by a hydrofluoric acid treatment. Graphene sheet of ~ 1×0.4 cm^2 was deposited on the wafer soon after to reduce the chance of oxide formation on the Si surface. Graphene acts as the anode of the Gr/Si junction and top electrode of the Gr/SiO_2/Si Metal-Oxide-Semiconductor (MOS) capacitor, formed at two sides of the junction [12, 13]. The second type of device, shown in Fig. 1(b), was made from a degenerately doped (~ 10^{18} cm^{-3}) n-type Si wafer. We used lithography and Reactive-Ion Etching (RIE) to fabricate a matrix of nanotips with ~ 450 nm height, ~ 50 nm top-diameter and 1.4 µm distance from each other (from here on, we will refer to the nanotip array as the Si-tips). Subsequently, the Si-tips were coated with a layer of SiO_2 deposited by CVD, then, the SiO_2 thickness was reduced through a chemical-mechanical planarization step until the Si-tips were revealed. Before the graphene transfer, the Si-tips were dipped in a 0.5% hydrofluoric acid solution for 10 s to remove the native SiO_2 from their apices [14]. In both devices graphene was transferred from commercial Cu foils onto the Si substrate using the wet transfer process [15]. Ohmic metal contact to graphene was fabricated by evaporation of Ti/Au metal stack through a shadow mask. In the first device, the cathode was established by depositing Ag paste on the exposed Si area of the wafer after scratching it to ensure ohmic contact. In the second device, Ag paste was deposited on the bottom of the substrate after the SiO_2 was removed by scratching. The I-V characterization was performed in vacuum (~10 mbar) using a Janis ST-500 micromanipulated probe station connected to a Keithley Semiconductor Characterization System 4200 (SCS-4200). The forcing bias was applied to the Ti/Au contact while the Ag contact was grounded. The responsivity was investigated using the light from a white LED array with controllable intensity up to 5 mW/cm^2 and spectral range 420 − 720 nm with two peaks at 454 nm and 536 nm, respectively.

3. Results and discussion

In Fig. 1(c) we show the current-voltage characteristics of the devices of Fig. 1(a) and (b), which we refer to as the flat Gr/Si and the Gr/Si-tip junction, respectively. Both devices were measured at 300 K and show a rectification behaviour with the forward current at positive bias typical of p-type graphene on n-type Si. At the bias of ∓1 V, the flat Gr/Si device displays a rectification factor of 10^3 while the Gr/Si-tip junction has a rectification ratio of 10^2 .

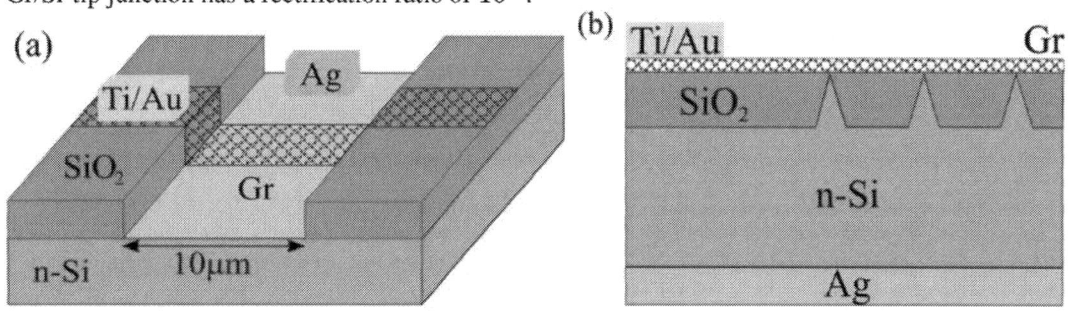

IOP Conf. Series: Journal of Physics: Conf. Series **956** (2017) 012019 doi:10.1088/1742-6596/956/1/012019

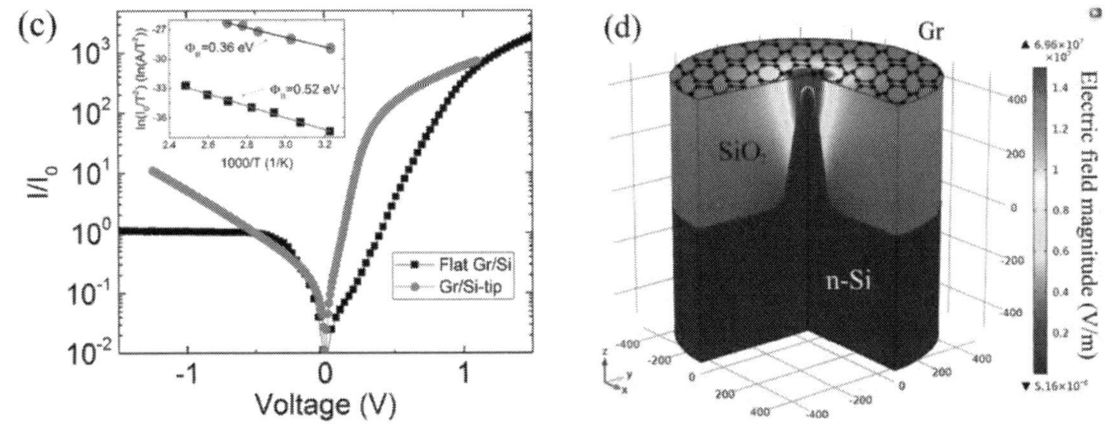

Figure 1. (a) Schematic view of the flat Gr/Si and (b) Gr/Si-tip devices. (c) I-V characteristics of the two devices, measured at 300 K and from -1.5 V to 1.5 V. The inset shows the Richardson plot of both devices where I_0 is the zero bias saturation current, obtained by extrapolating the forward current to 0 V. (d) Numerical simulation (by COMSOL Software) of the electric field around the tip at a reverse bias of -1 V. The intensity of the electric field is shown through a scale of colors in which blue indicates the minimum value while the red is the maximum. The Gr/Si junction in this simulation is considered like an ideal Schottky contact with a work function of 4.6 eV and a Schottky barrier height of 0.19 eV. The Gr/Si is at -1 V respect to the bottom n-Si substrate, which is grounded.

The I-V characteristic can be studied using the thermionic emission model:

$$I = I_0 \left[e^{\frac{qV}{nkT}} - 1 \right] \tag{1}$$

where

$$I_0 = AA^* \, T^2 \, e^{-\frac{\Phi_B}{kT}} \tag{2}$$

is the reverse current, A is the contact surface, $A^* = 4\pi q m^* k^2/h^3$ is the effective Richardson constant which is specific of the semiconductor type (for n-Si, $A^* = 120\,A/(K \cdot cm)^2$), m^* is the effective electron mass, q is the electronic charge, Φ_B (eV) is the Schottky barrier height, k is the Boltzmann constant and n is the ideality factor. The ideality factor is a phenomenological parameter that takes into account deviations from the thermionic regime, as carrier generation/recombination or effects due to the presence of oxide or other inhomogeneities at the Gr/Si interface. Rewriting equation (1) in the form:

$$\ln\left(\frac{I_0}{T^2}\right) = \ln(AA^*) - \frac{\Phi_B}{k}\frac{1}{T} \tag{3}$$

Φ_B and A^* can be evaluated from the slope and the intercept of the plot of $\ln(I_0\,/\,T^2)$ vs. $10^3/T$, respectively. In Tab. 1 we report the Schottky barrier height Φ_B, the ideality factor n and the Richardson constant A^* of the two devices.

Table 1. Schottky parameter of both devices

	Φ_B(eV)	n	A^* ($Acm^{-2}K^{-2}$)
Flat Gr/Si device	0.52	3.9	4.78×10^{-5}
Gr/Si-tip device	0.36	1.5	0.002

The difference in the barrier height can be ascribed to the stronger force image barrier lowering which depends on the forth root of the doping level, $\Delta\Phi_B \sim \sqrt[4]{N_D}$. The high value of the ideality factor n and the resulting A^*, which is lower than the theoretical value in metal-semiconductor junctions [13, 14, 16,

17], suggest the presence of an insulating layer and of inhomogeneity at the Gr/Si interface, which are higher for the flat Gr/Si device, due to the larger junction area.

Fig. 1(c) shows also that the reverse current of the flat Gr/Si device is constant while the one of the Gr/Si-tip junction increases exponentially with the bias. The reason for this difference lies in the geometry of the substrate. The electric field of the flat Gr/Si junction is homogeneous throughout the contact surface and weakly dependent on the applied bias. In the Gr/Si-tip device, the tip geometry greatly enhances the electric field (Fig. 1(d)), and the effect of bias on graphene becomes more pronounced. A change of the applied bias causes a shift of the graphene Fermi level and therefore affects the Schottky barrier height [2, 14], as shown in Fig. 2(a). The growing reverse saturation current of the Gr/Si-Tip device can be explained by a decrease of the Schottky barrier height, as confirmed experimentally in Fig. 2(b), where the barrier height has been estimated from equation (3) at different biases.

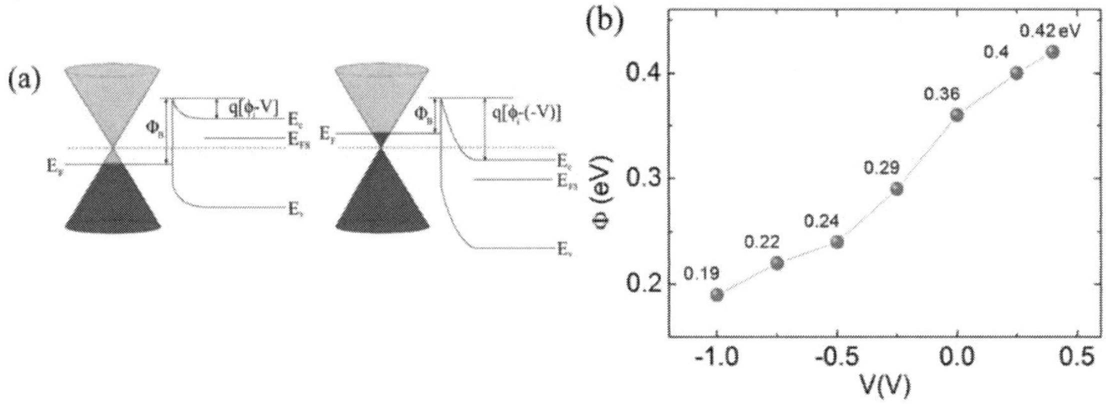

Figure 2. (a). Band diagram of the Gr/Si junction, in forward and in reverse bias. E_F is the graphene Fermi level, E_{FS} is the Si Fermi energy, E_c and E_v are the bottom and the top of the conduction and valence bands. ϕ_i is the build-in potential and V is the applied bias. In forward (reverse) bias, the potential shifts down (up) the graphene Fermi level increasing (decreasing) the Schottky barrier height. (b) Schottky barrier height versus bias, extracted using equation (3) on a series of I-V measurements at different temperatures (Ref. [12, 14] for details).

We also studied the responsivity of both devices by shining light from the top, on the graphene layer. Fig. 3(a) and (b) show the I-V characteristics of the two devices illuminated with white LED light at different illumination levels. The responsivity is defined as:

$$\mathcal{R} = \frac{I_{light} - I_{dark}}{P_{in}} \tag{4}$$

where I_{light} and I_{dark} are the reverse currents measured under illumination and in dark, P_{in} is the incident optical power. The flat Gr/Si device showed a responsivity linearly dependent on the incident power and reached 2.5 A/W at $V = -2.4$ V under the light intensity of 5 $\mathrm{mW/cm^2}$ [12], while the Gr/Si-tip junction achieved a responsivity of 3 A/W at $V = -0.5$ V under light intensity of 3$\mathrm{mW/cm^2}$ (this latter value is obtained considering the effective junction area, i.e. the total Gr/Si-tips contact area) [14]. Fig. 3(c) and (d) show the ON/OFF sequence of both devices at $V = -2.4$ V and $V = -0.5$ V, respectively. In both cases, the rising/falling times are limited by the measurement rate of our source measurement unit (2Hz). The higher responsivity of the Gr/Si-tip device points toward of a more efficient light collection as well as to an intrinsic photocharge avalanche multiplication by impact ionization caused by the enhanced electric field at the top of the tips.

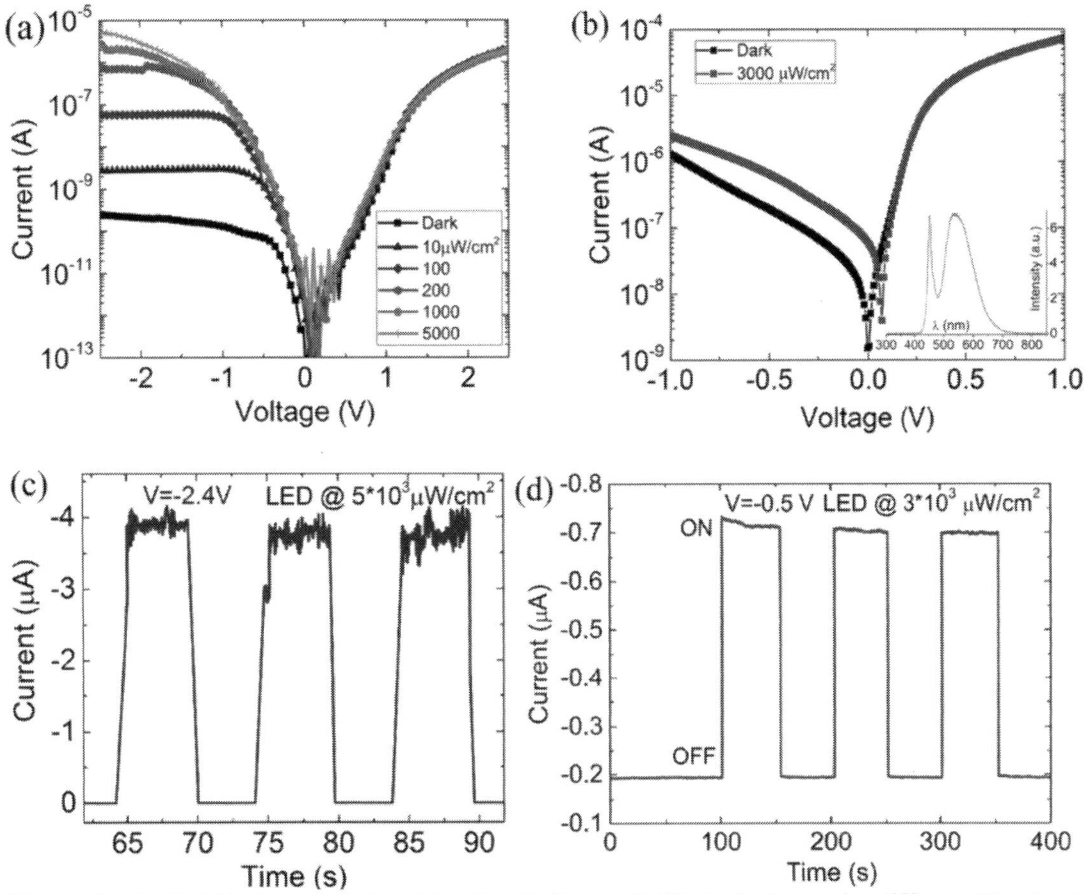

Figure 3. (a)-(b) I-V characteristic of the flat Gr/Si and Gr/Si-tip device under different illumination levels from a white LED array, whose spectrum is shown as inset of figure (b). (c)-(d) Dynamic measurement of the photocurrent at reverse bias $V = -2.4$ V for the flat Gr/Si device and $V = -0.5$ V for the Gr/Si-tip device.

Remarkably, the flat Gr/Si device displays a reverse current greater than the forward at maximum illumination level. We explain this phenomenon as due to Gr/SiO_2/Si MOS capacitor in parallel with the Gr/Si Schottky diode. In negative bias, the MOS capacitor accumulates photogenerated minority carries (holes), which diffuse to the Gr/Si junction region and contribute to reverse saturation current [12, 13, 18].

4. Conclusion

In conclusion, we have fabricated two types of devices with different geometries and measured their current-voltage characteristics and responsivity to white LED light. We have extracted the Schottky parameters and compared them to each other. These Schottky barrier heights of Gr/Si Flat device is in agreement with other Gr/Si heterojunction with flat geometry, while the Schottky barrier height of the Gr/Si-tip device is slightly lower due to the mentioned barrier lowering effect. Both device show a responsivity of one order of magnitude higher Ref. [11, 18], due to the embedded MOS capacitor and the nano-tip geometry, respectively. We have highlighted some advantages of the Gr/Si-tip approach, such as the higher photoresponse due to improved light collection and internal gain, as well as the possibility of tuning the Schottky barrier height. For the Gr/Si flat junction we have shown that high responsivity can be achieved by exploiting the parallel MOS capacitor, which is naturally built together with the junction. Our study clearly shows the importance of the substrate geometry, which can be used to enable specific functionalities of the Gr/Si junction.

5. References

[1] Sze S M and Ng K K 1969 *Physics of Semiconductor Devices 3rd edition* Wiley Interscience John Wiley & sons Inc. Hoboken New Jersey USA 2017 pp 158-169
[2] Di Bartolomeo A 2016 *Physics Reports* **606** 1–58
[3] Di Bartolomeo A, Giubileo F, Iemmo L, Romeo F, Russo S, Unal S, Passacantando M, Grossi V and Cucolo A M 2016 *Appl. Phys. Lett.* **109** 023510
[4] Giubileo F, Di Bartolomeo A, Martucciello N, Romeo F, Iemmo L, Romano P and Passacantando M 2016 *Nanomaterials* **6**(11) 206
[5] Pop E, Varshney V and Roy A 2012 *MRS Bull.* **37** 1273
[6] Castro Neto A H, Guinea F, Peres N, Novoselov K and Geim A 2009 *Rev. Modern Phys.* **81** 109-161
[7] Di Bartolomeo A, Santandrea S, Giubileo F, Romeo F, Petrosino M, Citro R, Barbara P, Lupina G, Schroeder T and Rubino A 2013 *Diamond and Related Materials* **38** 19–23
[8] Di Bartolomeo A, Giubileo F, Romeo F, Sabatino P, Carapella G, Iemmo L, Schroeder T and Lupina G 2015 *Nanotechnology* **26** 475202
[9] Giubileo F and Di Bartolomeo A 2017 *Progress in Surface Science* **92** 143–175
[10] Xia F, Perebeinos V, Lin Y and Wu Avouris Y 2011 *Nat. Nanotechnol.* **6** 179-184
[11] Selvi H, Unsuree N, Whittaker E, Halsall M P, Hill E W, Parkinson P and Echtermeyer T J 2017 *Arxiv* https://arxiv.org/abs/1706.09042
[12] Di Bartolomeo A, Luongo G, Giubileo F, Funicello N, Niu G, Schroeder T, Lisker M and Lupina G 2017 *2D Materials* **4** 025075
[13] Luongo G, Giubileo F, Genovese L, Iemmo L, Martucciello N and Di Bartolomeo A 2017 *Nanomaterials* **7** 158
[14] Di Bartolomeo A, Giubileo F, Luongo G, Iemmo L, Martucciello N, Niu G, Fraschke M, Skibitzki O, Schroeder T and Lupina G 2017 *2D Materials* **4** 015024
[15] Lupina G, Kitzmann J, Costina I, Lukosius M, Wenger C, Wolff A, Vaziri S, Östling M, Pasternak I, Krajewska A, Strupinski W, Kataria S, Gahoi A, Lemme M, Ruhl G, Zoth G, Luxenhofer O and Mehr W 2015 *ACS Nano* **9** 4776
[16] Liang S J, Hu W, Di Bartolomeo A, Adam S and Ang L K 2017 *Technical Digest - International Electron Devices Meeting IEDM 2016* 7838416 14.4.1-14.4.4 DOI: 10.1109/IEDM.2016.7838416
[17] Varonides A 2016 *Phys. Status Solidi C* **13** 1040-44
[18] Riazimehr S, Kataria S, Bornemann R, Haring Bolívar P, Ruiz F J G, Engström O, Godoy A and Lemme M C 2017 *ACS Photonics* **4** 1506–14

Acknowledgments

We acknowledge the economic support of Regione Campania through the projects POR FSE 2014-2020, Asse III Ob. Specifico l4, Avviso pubblico decreto dirigenziale n. 80 del 31/05/2016 and L.R. num. 5/2002 Finanziamento progetti annualità 2008, Prot. 2014, 0293185, 24/04/2014.

IOP Conf. Series: Journal of Physics: Conf. Series **956** (2017) 012020 doi:10.1088/1742-6596/956/1/012020

Inclusions of Si-atoms in Graphene nanostructures: a computational study on the ground-state electronic properties of Coronene and Ovalene

P Mocci, R Cardia and G Cappellini

Department of Physics, University of Cagliari, S.P. Monserrato-Sestu Km 0,700 I-09042 Monserrato (CA), ITALY

E-mail: paola.mocci@dsf.unica.it

Abstract. We present a computational study on graphene nanostructures, focusing our attention on Coronene ($C_{24}H_{12}$) and Ovalene ($C_{32}H_{14}$) molecules and their Si-atoms substituted counterparts. We used density functional theory to quantify the effects of the chemical modifications. For all the modified geometries (with single silicon atoms, dimers and trimers) we record an increase of the energy stability. Concerning the structural properties, the effect caused by Si-trimers is, in some configurations, for both the molecules under study, a deformation of the original structure associated with a loss of the flatness, typical of the pure systems.

1. Introduction

Polycyclic aromatic hydrocarbons (PAHs) in their crystalline and thin-film state are widely used as active elements in several opto-electronic devices, such as organic thin-film field-effect transistors, light-emitting diodes, photovoltaic cells, and liquid crystals [1]. Among them, the Circumacenes (Coronene, Ovalene, Circumanthracene, Circumtetracene, and Circumpentacene) have recently attracted attention [2, 3]. Coronene is made by the union of six benzene rings (with chemical formula $C_{24}H_{12}$), while Ovalene ($C_{32}H_{14}$) is obtained from Coronene by the addition of three benzene rings. The first one occurs naturally as the very rare mineral Carpathite, which is characterized by flakes of pure Coronene embedded in sedimentary rock and it can be tractable through hydrocracking processes [4]; the latter is a reddish-orange compound that can be synthesized, as Coronene, via the hydrocracking process of petroleum refining.
When compared with polymers, small molecules offer several advantages, such as the possibility to be easily purified by various techniques and processed both by evaporation and from solution [5, 6]. These molecules are here considered for their interesting properties: they are the smallest in size among the Circumacenes, but they qualify as nanometric systems and they are moreover characterized by a symmetric structure with a planar geometry.
These systems are of strong interest both for fundamental research (e.g. in astrochemistry and in condensed matter) and for solid state applications. Our choice to deal with this set of PAHs is mainly justifiable by the following reasons: (i) they present both a rigid backbone and an energy gap of moderate size rapidly converging with the size of the molecule [7]; (ii) their molecular packing in thin films might extremely differ from the classical herringbone structure found for the corresponding (oligo)acenes and thus might provide more efficient pathways for charge transport;

and (iii) the synthesis of Circumanthracene [8] as well as other versatile related molecules [9] has been successfully reported in the literature. Hence, these low-dimensional carbon structures could be considered as promising candidates for new emergent technologies [10].
The aim of this work consists in the study of the effects that silicon-atoms substitutions induce in Coronene and in Ovalene molecules with respect to their ground-state properties. This work follows other previous studies related to inclusion of Si atoms in Graphene [11, 12]. We consider initially one atom substitutions, double atoms substitutions (dimers) and triple substitutions (trimers) of Si, analyzing their effects on the morphological and electronic properties, using the Density Functional Theory (DFT)[13]. We will compare our results with existing data after previous theoretical and experimental works and concerning the ground-state electronic properties we will try to find possible trends.

2. Computational Methods

Following previous studies [2, 3], we performed geometry optimizations using the hybrid exchange-correlation functional B3LYP [14, 15, 16], in combination with the $6-31G^*$ basis-set, a valence double-ζ set augmented with d polarization functions for each atom. We used the B3LYP XC potential, since with respect to other possible choices (e.g. the PBE (Perdew-Burke-Ernzerhof)[17]), it has been proved to reproduce for different families of PAHs molecules better results, either for ground-state and the excited one [2, 18, 19, 20]. All the molecular relaxations have been performed without symmetry constraints. Geometry optimizations were obtained using tight convergence criteria, specified by maximum and root mean square gradient thresholds of $1.5 \cdot 10^{-5}$ and $1.0 \cdot 10^{-5}$ atomic units, respectively, and maximum and root mean square thresholds of the Cartesian step respectively of $6.0 \cdot 10^{-5}$ and $4.0 \cdot 10^{-5}$ atomic units. We employed NWChem [21], that is an all-electron gaussian-based computational package, to evaluate the general properties of Coronene and Ovalene substitiitutional clusters. The code is based on a numerical resolution of the Khon-Sham equations after a DFT Hamiltonian [13]. The exchange and correlation effects are treated using a particular functional entering in the Hamiltonian after the B3LYP scheme [14, 15, 16].

3. Si-atoms Insertions

According to the molecular symmetries, the possible non-equivalent insertions of silicon atoms are the following (see Fig. 1):

- 6 (9) configurations after single Si atom substitutions in Coronene (Ovalene);
- 4 (9) configurations after Si-dimer substitutions in Coronene (Ovalene);
- 4 (10) configurations after Si-trimer substitutions in Coronene (Ovalene).

We have found that, in the case of Coronene, the configurations number is reduced from 6 to 3, because of the equivalence between A-F, B-E and C-D atoms. On the other hand, considering the Ovalene structure, the addition of three benzene rings increases the possible configurations, entailing a gradual loss of the molecular symmetry. The configurations of substituted species were found to preserve the planar geometry of their parent molecules, after single and double Si-atoms inclusions. On the contrary, for both Coronene and Ovalene, the insertion of a trimer, in some case, breaks their planar symmetry (see fig. 3). This effect appears to be more pronounced for Ovalene than for Coronene as shown by the variation in interatomic distances connecting Si-C atoms. On the contrary, we verified that single and dimer substitutions, as for Coronene and Ovalene, preserve the flatness typical of the original systems: the deformations are planar and only localized.

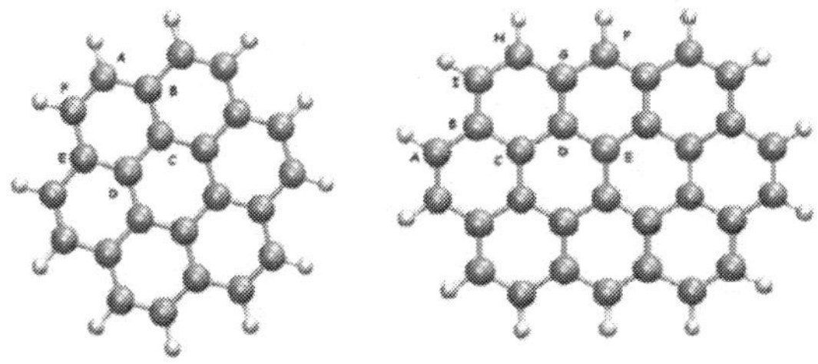

Figure 1. Coronene and Ovalene molecules. The Carbon atoms selected to be replaced by Silicon atoms are indicated by letters from A to F for Coronene (left), while from A to I for Ovalene (right). Dimers and trimers will be composed by the union of the adjacent single ones.

Table 1. Unsubstituted-systems ground-state total energies (E_0^n) and ground-state total energy differences $E_0^s - E_0^n$ (see equation 1). The percentage deviation Δ is also reported.

	Total Energy Differences (keV)	Δ (%)
Coronene	$(E_0^n = -25.1)$	–
single-substitution	-6.837	27.3
dimer-substitution	-14.678	58.5
trimer-substitution	-20.518	81.2
Ovalene	$(E_0^n = -33.4)$	–
single-substitution	-6.838	20.5
dimer-substitution	-13.678	40.9
trimer-substitution	-20.518	61.4

4. Ground-state Total Energies

Tab. 1 reports the computed data of the ground-state energy (E_0^n) for Coronene and Ovalene pure neutral systems and their substituted counterparts. In particular, the first column shows the energy differences between the original molecules and each substituted one. Because of the large configurations number, we have selected for each insertion (single, double and trimer) the most energetically stable structure among all the possibilities, to make a comparison with its corresponding pure parent. This comparison can be evaluated calculating the percentage change (Δ) passing from the pure system to the substituted one for each type of insertion, namely:

$$\Delta = \frac{E_0^s - E_0^n}{E_0^n} \tag{1}$$

where E_0^s and E_0^n are respectively the ground-state total energies for the substitued system and the unsubstituted one. We note that, from the energetic point of view, all Ovalene configurations show deeper energies than those of Coronene (that is an increase, in absolute value, of the ground-state total energy, E_0^n) and this is a consequence of the increase in the number of electrons, while the (Δ) variations are lower than in the case of Coronene. In particular, the most energetically favourable configurations are, for Coronene (Ovalene) the following ones: the single substitution corresponding to the insertion indicated by letter C (H); for the double substitution, the dimer composed by letter A-F (H-I); for the triple insertion, the trimer consisting of F-A-B (H-I-B). In general, the energy stabilty tends to descrease from the external to the inner molecular region:

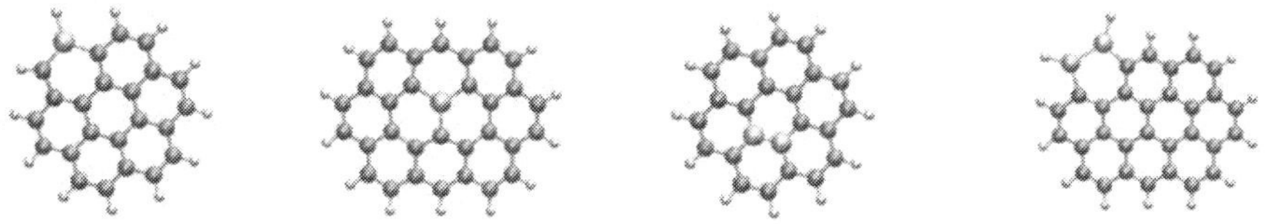

Figure 2. Coronene (a, c) and Ovalene (b, d) with single Si-atom inclusions and Si-dimers ones.

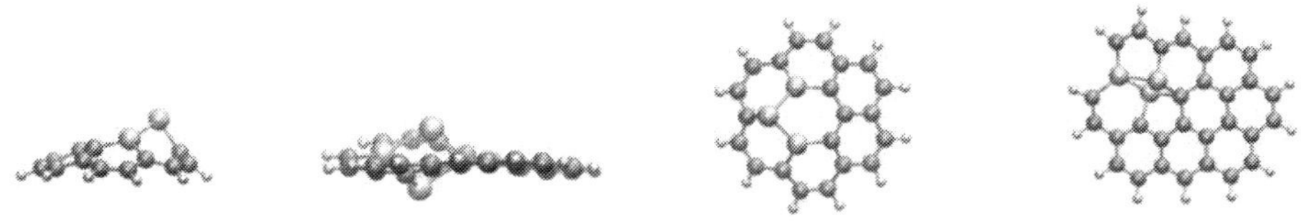

Figure 3. Coronene (a,c) and Ovalene (b,d) with a Si-trimer substitution placed in an internal molecular region.

this means that, in general, the most stable configurations are those in which the substitutional site is placed in the peripheral molecular areas, rather than in the inner part of the molecules (the only exception is the sigle susbistuted Coronene).

5. Morphological Properties

In this section we select an example for each species to show the effects induced on the structural properties by the Si-atoms insertions and what are the consequences as a function of the numbers of C-atoms replaced. In particular, we focus our attention on the trimer substitution for both the molecules. In fact, this modified configuration largely loses the flatness typical of the pure Circumacenes (see fig. 3). For Coronene, we selected the configuration in which the trimer substitution is made on the central ring (corresponding Si atoms are represented in yellow and indicated by numbers 1,2,3). For Ovalene the modified trimer is composed by the atoms labeled by B, C, D. Interatomic bond lenghts are shown in Table 2, where Si-C, Si-Si, C-C atomic distances are reported for these two specific configurations, labeling the selected atoms as indicated by fig. 4. Also the percentage changes of interatomic distance between C-C atoms that compose the parent molecules are shown (the medium value of C-C bond lenght for the unsubstituted compounds ($C - C_{unp}$) is 1.43 $\mathring{A}$). We analyzed quantitatively how the distances in the ring in which Silicon atoms are inserted have been modified in comparison to the original structures. In the case of trimer substitutions, it is possible to see that the distorsions are consistent and they concern the global structure, that loses noticeably its planar appearance. On the contrary, in the case of some single and double substitutions, after the present calculations, minor deformations of the rigid structure take place, but in general, for both Coronene and Ovalene, the deformations are only local and they determine a change of the interatomic bond lenghts, without any loss of planar feature.

Table 2. Interatomic bond lenghts between C-Si, Si-Si and C-C atoms for the two selected substituted molecules with a trimer. The percentage changes (Δ) with respect to the C-C unperturbed distance ($C-C_{unp}$) are also shown.

Coronene	Å	Δ(%)	**Ovalene**	Å	Δ(%)
$C-C_{unp}$	1.43	–	$C-C_{unp}$	1.43	–
C^1-Si^1	1.78	19.7	C^1-Si^B	1.84	22.3
C^2-Si^1	1.80	25.9	C^2-Si^B	1.84	22.3
C^5-Si^2	1.92	25.5	C^4-Si^D	1.92	25.5
C^6-Si^3	1.80	25.9	C^6-Si^C	1.92	25.5
C^4-Si^3	1.78	19.7	C^5-Si^D	2.04	29.9
Si^1-Si^2	2.22	35.6	Si^B-Si^C	2.68	46.6
Si^2-Si^3	2.22	35.6	Si^B-Si^D	2.69	46.8
C^2-C^3	1.45	1.37	C^2-C^3	1.37	-4
C^3-C^6	1.45	1.37	C^3-C^4	1.45	1.38
			C^4-C^7	1.37	-4

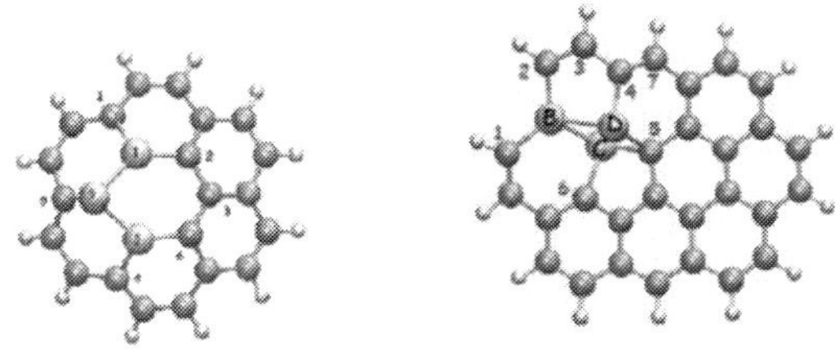

Figure 4. Coronene (a) and Ovalene (b) with a Si-trimer substitution in which Si and C atoms are labeled by letters and numbers, as reported in Table 2.

In particular, modifications imply that for Coronene the percentage change of the C-Si bond with respect to the corresponding C-C bond of the unsubstituted system goes from $\sim$ 20% to 26%. Likewise, a similar situation is recorded for Ovalene, where the parcentage change of the C-Si bond in comparison with the C-C one results between the range of 22% – 30%. Moreover, the distance connecting two silicon atoms (Si-Si bond) is larger in the case of Ovalene (with an increase of about $\sim$ 47% compared to the C-C bond) if compared to that of Coronene ($\sim$ 36%), confirming a greater deformation induced by the insertion in Ovalene molecule. On the contrary, the lenght of the C-C bond, in the case $C^2-C^3$and C^4-C^7, in the substituted Ovalene is subjected to a reduction of about $\sim$ 4% compared to the pure geometry, while for Coronene the C-C bond for C^2-C^3 and C^3-C^6 show a negligible variation of about $\sim$ 1.4%. This effect proves that the chemical modification implies different results depending on the position in which it has been made as well as on the system under study. After the here presented results for Si-atoms insertions the here considered isolated molecules seem to offer a different scenario in comparison to the infinite Graphene sheet [11, 12]. This point therefore deserves additional focused experimental studies associated with other theoretical ones to be fully understood.

Conclusions

We presented a systematic comparative study of Coronene and Ovalene in their unsubstituted and substituted forms, with Silicon single atoms, dimers and trimers insertions. In the

framework of DFT calculations, we found larger values of ground-state energies compared to the original structures, in both Coronene and Ovalene molecules as a consequence of the chemical modification.
Dealing with morphological properties, we observed that some trimer insertions (as for Coronene and Ovalene) are the substitutions which deform consistently the molecule in the out-of-plane directions. Other calculations, performed with different substitutional sites, have shown different behaviours in terms of out-of-plane distortions: for example, some peripheral substitutions did not present this effect. We expect that as the number of Si atoms substituted increases, there would be larger or morphologically different molecular distortions. The present systematic quantitive evaluation of the effect of Si-atoms substitutions on the structural properties of the two first members of the Circumacenes can be considered the preliminar step for future studies on the electronic (ground and excited) states and the optical properties of these systems.

Acknowledgments

P. M. gratefully acknowledges Sardinia Regional Government for the financial support of her PhD scolarship (P.O.R.Sardegna F.S.E. Operational Programme of the Autonomous Region of Sardegna, European Social Fund 2007-2013 Axis IV Human Resources, Objective I.3, Line of Activity I.3.1). G.C. and R.C. acknowledge partial financial support from IDEA-AISBL Bruxelles. G.C. acknowledges, moreover, partial financial support from "Progetto biennale d'Ateneo UniCa/FdS/RAS (Legge Regionale 7 agosto 2007, n. 7– annualita'; 2016)- Multiphysics theoretical approach to thermoelectricity". The authors acknowledge computational support by CINECA through ISCRA Initiative (Project SIGREOP).

References

[1] Perez-Jimenez A J and Sancho-Garcia J-C 2009 *IOP Pub. Ltd Nano.* **20** 47
[2] Malloci G, Cappellini G, Mulas G and Mattoni A 2011 *Chem. Phys.* **384** 19-27
[3] Cardia R, Malloci G, Cappellini G and Serra G 2016 Proc. of *SPIE* **9895** 302-308
[4] Jennings E, Montgomery W and Lerch P 2010 *J. Phys. Chem.* B **114** 48
[5] Anthony J E 2006 *Chem. Rev.* **106** 5028-5048
[6] Burke K B *et al* 2012 *Cryst. Growth Des* **12** 725-731
[7] Jiang D and Dai S 2008 *Chem. Phys. Lett.* **466** 72
[8] Broene R D and Diederich F 1991 *Tetrah. Letters* **32** 5227
[9] Wu J, Pisula W and Mullen K 2007 *Chem. Rev.* **107** 718
[10] Sancho-Garcia J C and Prez-Jimnez A J 2009 *Phys. Chem. Chem. Phys.* **11** 2741-2746
[11] Lopez-Bezanilla A, Zhou W and Idrobo J C 2014 *The J. Phys. Chem. Lett.* **5** 1711-1718
[12] Zhou W *et al* 2012 *Phys. Rev. Letters* **109** 206803
[13] Kohn W 1999 *Rev. Mod. Phys.* **71** 1253
[14] Becke A D, 1993 *J. Chem. Phys.* **98** 5648
[15] Lee C , Yang W, and Parr R G 1988 *Phys. Rev. B* **37** 785
[16] Stephens P J, Devlin P J, Chabalowski C F, and Frisch M J, 1994 *J. Phys. Chem.* **98** 11623
[17] Ernzerhof M *et al* 1999 *Journal of Chem. Phys.* **110** 11
[18] Cardia R, Malloci G, Mattoni A, and Cappellini G 2014 *J. Phys. Chem. A* **118** 5170–5177
[19] Cardia R *et al* 2016 *Phys. Rev. B* **93** 235132
[20] Cappellini G, Malloci G and Mulas G, 2009 *Superlattices and Microstruc.* **46** 14-18
[21] Valiev M *et al* 2010 *Comput. Phys. Commun.* **181** 1477

Non-thermal Unruh radiation for flavour neutrinos

M Blasone[1,2], **G Lambiase**[1,2] **and G G Luciano**[1,2]

[1] Dipartimento di Fisica, Università degli Studi di Salerno, Via Giovanni Paolo II, 132 I-84084 Fisciano (SA), Italy
[2] INFN Gruppo Collegato di Salerno - Sezione di Napoli c/o Dipartimento di Fisica, Università di Salerno, Italy

E-mail: gluciano@sa.infn.it

Abstract. In the quantum field theory framework, both flavour mixing and Unruh effect have been shown to arise from Bogoliubov transformations connecting inequivalent Hilbert spaces. In the present work, we study how these transformations combine when field mixing for an accelerated observer (Rindler observer) is considered. In particular, a simplified two–flavour model involving Dirac neutrino fields is analyzed. In spite of such basic setting, we find that the spectrum of Unruh radiation gets significantly modified, losing its characteristic thermal behaviour. Exploiting this result, the possibility of fixing new constraints on the neutrino squared mass differences is investigated.

1. Introduction

The topics of neutrino flavour mixing and oscillations have been increasingly investigated since Pontecorvo's pioneering idea [1]. Although the theoretical basis of neutrino mixing have so far been well-established [2] and experimental developments have successfully confirmed the occurrence of flavour oscillations [3], the full picture has not yet been revealed. Puzzling questions, indeed, are still open, such as the origin of neutrino masses in the Standard Model or the non-trivial condensate structure of the vacuum for mixed fields. The latter issue, in particular, has been widely investigated after the unitary inequivalence between mass and flavour vacua in quantum field theory (QFT) was pointed out [4], thereby showing the limits of the quantum mechanical (QM) approach in the analysis of neutrino mixing.

Up to now, the QFT formalism for mixed neutrinos has been developed only within Minkowski framework. Such an approach would be strictly workable only for solar, atmospheric and reactor neutrinos. In order to get valuable information on the structure of the early universe and the properties of super-large celestial objects, however, cosmological neutrinos need to be considered. Because of their passage in the vicinity of supermassive black holes and star clusters, for these particles a treatment in the context of QFT on curved background is required. A first step along this direction has been recently taken in Ref. [5], where the effects of gravity on mixing transformations have been simulated with a uniform acceleration for the case of bosonic fields.

In the present work, starting from these preliminary results, the QFT of two mixed neutrino fields is analyzed in a uniformly accelerated frame. Despite such a minimal setting, the Bogoliubov transformation arising from the Rindler spacetime structure is found to non-trivially combine with the one hiding in field mixing[1]. As a result, Unruh radiation loses its original

[1] Bogoliubov transformations are isomorphisms of the canonical (anti-)commutation relation algebra. In other

thermal nature, potentially providing new constraints on the neutrino squared mass differences.

The paper is organized as follows: in Sec. 2 we briefly review the second quantization of the free Dirac spinor field in Minkowski spacetime (MS). Sec. 3 is devoted to the discussion of the field quantization in Rindler spacetime (RS) and the related Unruh effect. The QFT of two mixed neutrinos is analyzed in Sec. 4, both for inertial and uniformly accelerated observers. Conclusions are shortly summarized in the last section.

Throughout all the work, the metric $\eta_{\mu\nu} = \mathrm{diag}(+1,-1,-1,-1)$ and natural units $\hbar = c = 1$ are used. Moreover, the following notation is adopted: $x = \{t, \mathbf{x}\}$, $\mathbf{x} = \{x^1, \vec{x}\}$, $\vec{x} = \{x^2, x^3\}$.

2. Field quantization in MS: plane–wave and boost–mode representations

In MS the standard plane–wave expansion of the free Dirac neutrino field reads

$$\psi(x) = \sum_{r=1,2} \int d^3k \left[a^r_{\mathbf{k}}\, \psi^{r+}_{\mathbf{k}}(x) + b^{r\dagger}_{\mathbf{k}}\, \psi^{r-}_{\mathbf{k}}(x) \right], \tag{1}$$

where $\psi^{r+}_{\mathbf{k}} = N_k u^r_{\mathbf{k}}\, e^{-ik\cdot x}$ and $\psi^{r-}_{\mathbf{k}} = N_k v^r_{\mathbf{k}}\, e^{+ik\cdot x}$ are the positive and negative frequency plane waves, respectively, with $\omega_k = \sqrt{m^2 + |\mathbf{k}|^2}$ and $N_k = (2\pi)^{-3/2}\sqrt{\frac{m}{\omega_k}}$ (see Ref. [6] for the convention used for the spinors $u^r_{\mathbf{k}}$ and $v^r_{\mathbf{k}}$). The $a^r_{\mathbf{k}}$ and $b^r_{\mathbf{k}}$ are the annihilation operators for the ordinary Minkowski vacuum: $a^r_{\mathbf{k}}|0\rangle_M = b^r_{\mathbf{k}}|0\rangle_M = 0$. They are assumed to be canonical: $\left\{a^r_{\mathbf{k}}, a^{s\dagger}_{\mathbf{k}'}\right\} = \left\{b^r_{\mathbf{k}}, b^{s\dagger}_{\mathbf{k}'}\right\} = \delta_{rs}\,\delta^3(k-k')$, with all other anticommutators vanishing.

In order to analyze the connection between the Minkowski and Rindler quantum constructions, an alternative quantization in MS comes in handy. Since the time evolution of the Rindler observer corresponds to an infinite succession of Minkowski boost transformations [7], we expand the spinor field as follows

$$\psi(x) = \sum_{j=1,2} \int d^3\kappa \left[c^j_{\kappa}\, \Psi^{j-}_{\kappa}(x) + d^{j\dagger}_{\kappa}\, \Psi^{j+}_{\kappa}(x) \right], \tag{2}$$

where

$$\Psi^{j\mp}_{\kappa}(x) = \frac{1}{2} N^{\mp}_{\kappa} \Bigg[X^j_k\, e^{\pm\frac{i\pi}{2}\left(i\Omega - \frac{1}{2}\right)} \int_{-\infty}^{+\infty} d\theta\, e^{i\mu_k[x^1 \sinh\theta \mp t\cosh\theta]}\, e^{\mp\left(i\Omega-\frac{1}{2}\right)\theta}$$
$$+ Y^j_k\, e^{\pm\frac{i\pi}{2}\left(i\Omega + \frac{1}{2}\right)} \int_{-\infty}^{+\infty} d\theta\, e^{i\mu_k[x^1 \sinh\theta \mp t\cosh\theta]}\, e^{\mp\left(i\Omega+\frac{1}{2}\right)\theta} \Bigg]\, e^{i\vec{k}\cdot\vec{x}}, \tag{3}$$

(the spinors X^j_k and Y^j_k are defined as in Ref. [7]). These modes are eigenfunctions of the Lorentz momentum operator M_{01}, with eigenvalue Ω. The normalization is defined by $N^{\mp}_{\kappa} = e^{\pm\frac{1}{2}\pi\Omega}/(2\pi\sqrt{\mu_k})$, where the subscript κ stands for $\kappa = \{\Omega, \vec{k}\}$.

It is not difficult to prove that the unfamiliar quantization in Eq. (2) is equivalent to the plane–wave construction above introduced. From Eq. (3), indeed, it arises that the boost–modes Ψ^{j-}_{κ} (Ψ^{j+}_{κ}) are linear combination of positive (negative) frequency plane waves alone, with $\omega_k = \mu_k\cosh\theta$ ($\omega_k = -\mu_k\cosh\theta$) and $k_1 = \mu_k\sinh\theta$. This means that the transformation connecting the two quantum schemes does not mix the positive and negative frequency modes. Projecting the spinor field in Eq. (1) on Ψ^{j-}_{κ} and exploiting the orthonormality of plane–waves with respect to the scalar product in MS, we thus obtain

$$c^j_{\kappa} = (\Psi^{j-}_{\kappa}, \psi)_M = \sum_{r=1,2} \int_0^{+\infty} dk_1\, F_{j,r}(k_1, \Omega)\, a^r_{\mathbf{k}}, \tag{4}$$

words, they are linear transformations of ladder operators that preserve the algebraic relations among them.

where

$$F_{j,r} = \frac{2\,\pi^2}{\omega_k}\,N_\kappa^-\,N_k\left[\left(\frac{\omega_k+k_1}{\omega_k-k_1}\right)^{i\frac{\Omega}{2}+\frac{1}{4}} e^{i\frac{\pi}{2}\left(i\Omega+\frac{1}{2}\right)}\,X_k^{j\dagger}\;+\;\left(\frac{\omega_k+k_1}{\omega_k-k_1}\right)^{i\frac{\Omega}{2}-\frac{1}{4}} e^{i\frac{\pi}{2}\left(i\Omega-\frac{1}{2}\right)}\,Y_k^{j\dagger}\right]u_{\mathbf{k}}^r$$

(a similar connection between the b and d operators can be derived by projecting the field on Ψ_κ^{j+}). From the previous relations, it arises that the vacuum annihilated by c_κ^j and $a_{\mathbf{k}}^r$ is the same. Moreover, by a straightforward calculation, it is possible to show that the transformation in Eq. (4) is canonical. These two statements enable us to definitively conclude that the quantum formalisms above introduced are equivalent. Therefore, we are allowed to use the latter in our derivation of Unruh effect.

3. Field quantization in RS: Unruh effect

Let us now turn to the problem of quantizing the Dirac spinor field in non–inertial frames. As known, the natural framework for describing the motion of an accelerated observer in flat spacetime is the Rindler background. The analysis of flavour mixing for such an observer thus requires a preliminary discussion about the causal structure induced by this metric.

In Minkowski coordinates, the line element has the well known form $ds^2 = (dt)^2 - (dx^1)^2 - \sum_{j=2,3}(dx^j)^2$. Under the coordinate transformation $t = \xi \sinh\eta$, $x^1 = \xi\cosh\eta$, it becomes[2]

$$ds^2 = (dt)^2 - (dx^1)^2 - \sum_{j=2,3}(dx^j)^2 \;\underset{Rindler\ coord.}{\longrightarrow}\; ds^2 = \xi^2 d\eta^2 - d\xi^2 - \sum_{j=2,3}(dx^j)^2 . \tag{5}$$

The trajectory of an observer moving with constant proper acceleration a along the x^1–axis is given by $\xi(\tau) = \text{const} \equiv a^{-1}$, $\vec{x}(\tau) = \text{const}$, where τ is the proper time measured along the line. From Eq. (5), it thus follows that a Rindler observer in the right wedge $R = \{x|x^1 > |t|\}$ is causally disconnected from one in the left sector $L = \{x|x^1 < -|t|\}$. The null hyperplane $t = |x^1|$ $(t = -|x^1|)$, indeed, acts for him as future (past) event horizon.

Bearing this in mind, let us now solve the Dirac equation in the wedge R; similar calculations can be performed in the other three sectors of MS (see, for example, Refs. [7, 8]). Using the tetrad formalism, the Dirac equation in Rindler coordinates takes the form

$$i\partial_\eta\psi = \left(-i\xi\,\alpha^j\partial_j - \frac{1}{2}i\alpha^1 + m\xi\,\beta\right)\psi, \tag{6}$$

where $\alpha^j = \gamma^{\bar{0}}\gamma^{\bar{j}}$ and $\beta = \gamma^{\bar{0}}$. Here $\gamma^{\bar{\mu}}$ are the analogous in Rindler spacetime of the Dirac matrices γ^μ, satisfying the generalized condition $\gamma^{\bar{\mu}}\gamma^{\bar{\nu}} + \gamma^{\bar{\nu}}\gamma^{\bar{\mu}} = 2g^{\bar{\mu}\bar{\nu}}$ with $g^{\bar{\mu}\bar{\nu}}$ given in Eq. (5). We look for solutions of positive frequency with respect to the Rindler time η, in the form

$$\Psi_\kappa^{j\,R}(\eta,\xi,\vec{x}) = N_\Omega\left(X_k^j\,K_{i\Omega-\frac{1}{2}}(\mu_k\xi) + Y_k^j\,K_{i\Omega+\frac{1}{2}}(\mu_k\xi)\right)e^{-i\Omega\eta}e^{i\vec{k}\cdot\vec{x}}, \qquad j=1,2, \tag{7}$$

where $N_\Omega = \frac{1}{4\pi^2\sqrt{\mu_k}}\sqrt{\cosh(\pi\Omega)}$, $K_{i\Omega\pm\frac{1}{2}}$ is the modified Bessel function and $\kappa \equiv \{\Omega, \vec{k}\}$.

In order to analyze the relationship between the Minkowski and Rindler quantum constructions, it is useful to express the solutions of Eq. (7) in Minkowski coordinates. In this connection, let us observe that, after the coordinate transformation $(t, x^1) \to (\eta, \xi)$, the tetrads - and therefore the spinors - undergo an orthogonal rotation of the following form: $\Psi(t, x^1) = \exp\left(\frac{1}{2}\gamma^{\bar{0}}\gamma^{\bar{1}}\eta\right)\Psi(\eta,\xi)$. Eq. (7) thus becomes

$$\Psi_\kappa^{j\,R}(t,x^1,\vec{x}) = N_\Omega\left(X_k^j\,K_{i\Omega-\frac{1}{2}}(\mu_k\xi)\;e^{-\left(i\Omega-\frac{1}{2}\right)\eta} + Y_k^j\,K_{i\Omega+\frac{1}{2}}(\mu_k\xi)\,e^{-\left(i\Omega+\frac{1}{2}\right)\eta}\right)e^{i\vec{k}\cdot\vec{x}}. \tag{8}$$

[2] Notice that $\vec{x} \equiv \{x^2, x^3\}$ is common to both sets of coordinates.

It must be clear that this equation is still defined only in the right wedge R. However, as shown in Ref. [7], it could be linked to the solutions of the Dirac equation in the other sectors of MS by analytically continuing these functions across the horizons. The two possible paths of analytical continuation, one rotating clockwise from the region R to L, the other counterclockwise, correspond to the two different global representations of the boost eigenfunctions $\Psi_\kappa^{j\mp}$ in Eq. (3).

Holding all the necessary tools, the Unruh effect for the free spinor field can now be promptly derived. Following the standard procedure first used by Unruh [9], let us quantize the field in terms of solutions of the Dirac equation which are eigenfunctions of the Rindler Hamiltonian (that is to say, of the Lorentz boost generator in Rindler coordinates), orthonormal and analytic in the whole MS. The functions we are looking for are the following

$$\mathcal{R}_\kappa^j = \frac{1}{\sqrt{2\cosh(\pi\Omega)}}\left(e^{\frac{\pi\Omega}{2}}\,\Psi_\kappa^{j-} + e^{-\frac{\pi\Omega}{2}}\,\Psi_\kappa^{j+}\right), \quad \mathcal{L}_\kappa^j = \frac{1}{\sqrt{2\cosh(\pi\Omega)}}\left(e^{-\frac{\pi\Omega}{2}}\,\Psi_\kappa^{j-} - e^{\frac{\pi\Omega}{2}}\,\Psi_\kappa^{j+}\right), \tag{9}$$

where $\Psi_\kappa^{j\pm}$ are the globally defined modes in Eq. (3). One can easily verify that the functions $\mathcal{R}_\kappa^j$ vanish in the left wedge, whereas they consistently reduce to the Rindler solutions of Eq. (8) in the right sector, thereby providing the correct global representation of the modes for a uniformly accelerated observer in RS (the inverse behaviour, of course, is manifested by the $\mathcal{L}_\kappa^j$).

Now, inverting Eq. (9) with respect to $\Psi_\kappa^{j\pm}$ and inserting into Eq. (2), we have

$$\psi(x) = \sum_{j=1,2}\int d^3\kappa\left[r_\kappa^j\,\mathcal{R}_\kappa^j + r_\kappa^{j\dagger}\,\mathcal{R}_{\tilde\kappa}^j + l_\kappa^j\,\mathcal{L}_{\tilde\kappa}^j + l_\kappa^{j\dagger}\,\mathcal{L}_\kappa^j\right], \tag{10}$$

where $\tilde\kappa \equiv \{-\Omega, \vec{k}\}$ and

$$r_\kappa^j = \frac{c_\kappa^j\,e^{\frac{\pi\Omega}{2}} + d_\kappa^{j\dagger}\,e^{-\frac{\pi\Omega}{2}}}{\sqrt{2\cosh(\pi\Omega)}}, \quad l_\kappa^{j\dagger} = \frac{c_\kappa^j\,e^{-\frac{\pi\Omega}{2}} - d_\kappa^{j\dagger}\,e^{\frac{\pi\Omega}{2}}}{\sqrt{2\cosh(\pi\Omega)}}. \tag{11}$$

The so obtained *thermal Bogoliubov transformations* allow us to determine the spectrum of Rindler quanta in the inertial vacuum. A straightforward calculation leads to

$${}_M\langle 0|r_\kappa^{i\dagger}\,r_{\kappa'}^j|0\rangle_M = {}_M\langle 0|l_\kappa^{i\dagger}\,l_{\kappa'}^j|0\rangle_M = \frac{1}{e^{2\pi\Omega}+1}\,\delta_{ij}\,\delta^3(\kappa-\kappa') \equiv \frac{1}{e^{\frac{a\Omega}{T}}+1}\,\delta_{ij}\,\delta^3(\kappa-\kappa'), \tag{12}$$

where in the last step we have taken into account the fact that the proper energy of the quanta detected by a Rindler observer with acceleration a is $a\Omega$, rather than Ω. From Eq. (12) it thus follows that the vacuum radiation perceived by the Rindler observer has a thermal spectrum, according to Fermi–Dirac statistics with temperature $T = \frac{a}{2\pi}$ [9].

4. QFT of mixed fields: inertial and accelerated frames

In a simplified two–flavour model, mixing transformations for Dirac neutrino fields read

$$\psi_e(x) = \psi_1(x)\cos\theta + \psi_2(x)\sin\theta, \qquad \psi_\mu(x) = -\psi_1(x)\sin\theta + \psi_2(x)\cos\theta. \tag{13}$$

Here ψ_χ, $\chi = e, \mu$ are the neutrino fields with definite flavours; ψ_i, $i = 1, 2$ are the free neutrino fields with definite masses m_i; θ is the mixing angle.

Exploiting the completeness of the plane-waves $\psi_{\mathbf{k},}^{r\pm}$, the following free field–like expansions for flavour fields in MS can be adopted [4]

$$\psi_\ell(x) = \sum_{r=1,2}\int d^3k\,e^{i\mathbf{k}\cdot\mathbf{x}}\left[a_{\mathbf{k},\ell}^r(t)\,u_{\mathbf{k},\sigma}^r(t) + b_{-\mathbf{k},\ell}^{r\dagger}(t)\,v_{-\mathbf{k},\sigma}^r(t)\right], \quad (\ell,\sigma) = (e,1),(\mu,2), \tag{14}$$

where the notation: $u^r_{\mathbf{k},\sigma}(t) \equiv u^r_{\mathbf{k},\sigma}\, e^{-i\omega_k t}$, $v^r_{-\mathbf{k},\sigma}(t) \equiv v^r_{-\mathbf{k},\sigma}\, e^{+i\omega_k t}$ has been introduced.

For field quanta moving along the same direction as the Rindler observer (i.e., in the reference frame such that $\mathbf{k} = (k, 0, 0)$), the *flavour operator* $a^r_{\mathbf{k},e}$ assumes the form

$$a^r_{\mathbf{k},e}(t) = \cos\theta\, a^r_{\mathbf{k},1} \,+\, \sin\theta\, \left(U^*_{\mathbf{k}}(t)\; a^r_{\mathbf{k},2} \,-\, V_{\mathbf{k}}(t)\; b^{-r\dagger}_{-\mathbf{k},2}\right) \tag{15}$$

(similar expressions hold for the other ladder operators in Eq. (14)), where the mixing Bogoliubov coefficients $U_{\mathbf{k}}$ and $V_{\mathbf{k}}$ are given by

$$U_{\mathbf{k}}(t) \,=\, \left(\frac{\omega_{k,1}+m_1}{2\omega_{k,1}}\right)^{\frac{1}{2}} \left(\frac{\omega_{k,2}+m_2}{2\omega_{k,2}}\right)^{\frac{1}{2}} \left(1+\frac{k^2}{(\omega_{k,1}+m_1)(\omega_{k,2}+m_2)}\right) e^{i(\omega_{k,2}-\omega_{k,1})\,t}, \tag{16}$$

$$V_{\mathbf{k}}(t) \,=\, \left(\frac{\omega_{k,1}+m_1}{2\omega_{k,1}}\right)^{\frac{1}{2}} \left(\frac{\omega_{k,2}+m_2}{2\omega_{k,2}}\right)^{\frac{1}{2}} \left(\frac{k}{\omega_{k,2}+m_2}-\frac{k}{\omega_{k,1}+m_1}\right) e^{i(\omega_{k,2}+\omega_{k,1})\,t}. \tag{17}$$

By use of these equations, one can easily prove that flavour operators, just like the corresponding ones for definite mass fields, satisfy the canonical anticommutation relations at equal times.

The foregoing discussion is completely within the usual Minkowski framework. Thus it arises the question about how the above formalism gets modified for an accelerated observer. For this purpose, using the same approach as in Sec. 2, let us quantize the mixed fields in Eq. (13) as follows

$$\psi_\ell(x) = \sum_{j=1,2}\int d^3\kappa\, \left[c^j_{\kappa,\ell}\, \Psi^{j-}_{\kappa,\sigma}(x) \,+\, d^{j\dagger}_{\kappa,\ell}\, \Psi^{j+}_{\kappa,\sigma}(x)\right], \quad (\ell,\sigma) = (e,1), (\mu,2), \tag{18}$$

where $\Psi^{j\mp}_{\kappa,\sigma}$ are the boost–modes in Eq. (3) and

$$c^j_{\kappa,\ell} = \sum_{r=1,2}\int_0^{+\infty} dk_1\, F_{j,r,\sigma}(k_1,\Omega)\, a^r_{\mathbf{k},\sigma}, \qquad (\ell,\sigma) = (e,1), (\mu,2) \tag{19}$$

as for free fields (see the corresponding transformation in Eq. (4)). Notice that for quanta such that $\mathbf{k} = (k, 0, 0)$, the subscript κ reduces to $\kappa = (\Omega, 0, 0)$.

Equation (18) provides a springboard for analyzing flavour mixing in accelerated frames. By extending the Unruh quantum construction in Eq. (10) to the mixed fields, indeed, we are immediately allowed to write down for them the following expansions

$$\psi_\ell(x) = \sum_{j=1,2}\int d^3\kappa\left[r^j_{\kappa,\ell}\,\mathcal{R}^j_{\kappa,\sigma} + r^{j\dagger}_{\kappa,\ell}\,\mathcal{R}^j_{\tilde\kappa,\sigma} + l^j_{\kappa,\ell}\,\mathcal{L}^j_{\tilde\kappa,\sigma} + l^{j\dagger}_{\kappa,\ell}\,\mathcal{L}^j_{\kappa,\sigma}\right], \quad (\ell,\sigma) = (e,1), (\mu,2), \tag{20}$$

where the flavour operators for the Rindler observer arise from the interplay between the mixing and thermal Bogoliubov transformations in Eqs. (11) and (19), respectively, i.e.

$$r^j_{\kappa,\ell} = \frac{1}{\sqrt{2\cosh(\pi\Omega)}}\sum_{r=1,2}\int_0^{+\infty} dk_1\left(e^{\frac{\pi\Omega}{2}}\, F_{j,r,\sigma}(\Omega,k_1)\, a^r_{\mathbf{k},\sigma} + e^{-\frac{\pi\Omega}{2}}\, G_{j,r,\sigma}(\Omega,k_1)\; b^{r\dagger}_{\mathbf{k},\sigma}\right), \tag{21}$$

with

$$G_{j,r} \,=\, \frac{2\,\pi^2}{\omega_k}\, N^+_\kappa\, N_k\left[\left(\frac{\omega_k+k_1}{\omega_k-k_1}\right)^{i\frac{\Omega}{2}+\frac{1}{4}} e^{i\frac{\pi}{2}(i\Omega+\frac{1}{2})}\; X^{j\dagger}_k \,+\, \left(\frac{\omega'+k'_1}{\omega'-k'_1}\right)^{i\frac{\Omega}{2}-\frac{1}{4}} e^{i\frac{\pi}{2}(i\Omega-\frac{1}{2})}\; Y^{j\dagger}_k\right] v^r_{\mathbf{k}}$$

(similarly for $l^{j}_{\kappa,\ell}$). In the simplest case of $t = \eta = 0$, the spectrum of mixed neutrinos in the inertial vacuum thus assumes the form

$$
{}_M\langle 0|r^{i\dagger}_{\kappa,\chi}\, r^{j}_{\kappa',\chi}|0\rangle_M = \frac{1}{e^{\frac{a\,\Omega}{T}}+1}\,\delta_{ij}\,\delta(\kappa-\kappa') + \sin^2\theta\left[\frac{e^{\frac{\pi\,(\Omega+\Omega')}{2}}}{2\sqrt{\cosh(\pi\Omega)\,\cosh(\pi\Omega')}}\,N^{ij}_{F,F}(\Omega,\Omega') - \frac{e^{-\frac{\pi\,(\Omega+\Omega')}{2}}}{2\sqrt{\cosh(\pi\Omega)\,\cosh(\pi\Omega')}}\,N^{ij}_{G,G}(\Omega,\Omega')\right]\delta_{ij}, \quad (22)
$$

where

$$
N^{ij}_{F,F} = \sum_{r=1,2}\int F^*_{i,r}(k_1,\Omega)\,F_{j,r}(k_1,\Omega')\,|V_{\mathbf{k}}|^2, \qquad N^{ij}_{G,G} = \sum_{r=1,2}\int G^*_{i,r}(k_1,\Omega)\,G_{j,r}(k_1,\Omega')\,|V_{\mathbf{k}}|^2.
$$

One can immediately verify that Eq. (22) consistently reduces to the standard Unruh result in Eq. (12) for $\theta \to 0$ and/or $m_1 = m_2$ (see Eq. (17)), as it must be in absence of mixing. In the realistic limit $\frac{\Delta m^2}{m^2} \equiv \frac{|m_2^2-m_1^2|}{m^2} \ll 1$, however, it can be further manipulated, thus giving [10]

$$
{}_M\langle 0|r^{i\dagger}_{\kappa,\chi}\, r^{j}_{\kappa',\chi}|0\rangle_M = \frac{1}{e^{\frac{a\,\Omega}{T}}+1}\,\delta_{ij}\,\delta(\kappa-\kappa') + \mathcal{O}\left(\frac{\Delta m^2}{m^2}\right). \quad (23)
$$

Therefore, for mass differences that are comparable with the acceleration, the vacuum radiation experienced by the Rindler observer gets non-trivially modified, resulting in the sum of the usual Unruh contribution plus non-thermal corrections arising from flavour mixing and depending only on the parameter $\Delta m^2/m^2$. Beyond the conceptual relevance of such a result, Eq. (23) can thus be exploited to fix new potential constraints on the neutrino squared mass differences.

An heuristic interpretation for the origin of the modified spectrum in Eq. (23) can be found in Ref. [5], where the same scenario has been analyzed for the case of bosonic mixed fields.

5. Conclusions

The QFT of two mixed neutrinos has been analyzed in an accelerated frame as a basis for investigating gravitational effects on mixing transformations. Despite this minimal setting, the Bogoliubov transformation arising from the Rindler spacetime structure has been found to suggestively combine with the one built in field mixing. As a result, Unruh radiation loses its thermality, thus providing an alternative way to constrain the neutrino squared mass differences.

We stress that the formalism developed in this work provides a useful tool for analyzing neutrino flavour oscillations in the context of QFT on curved background. Following what has been done in Ref. [4] within Minkowski framework, indeed, the oscillation probability formulas for an accelerated observer can be derived by defining the one flavour neutrino state and then performing similar calculations to those in Eq. (22). Work is in progress along this direction.

References

[1] Bilenky S M and Pontecorvo B 1978 *Phys. Rept.* **41** 225
[2] Giunti C and Kim C W 2007 *Fundamentals of Neutrino Physics and Astrophysics* Oxford UK: Univ. Pr.
[3] Fukuda Y *et al* [Super-Kamiokande Collaboration] 1998 *Phys. Rev. Lett.* **81** 1562; Fukuda S *et al* [Super-Kamiokande Collaboration] 2000 *Phys. Rev. Lett.* **85** 3999; Abe S *et al* [KamLAND Collaboration] 2008 *Phys. Rev. Lett.* **100** 221803
[4] Blasone M and Vitiello G 1995 *Annals Phys.* **244** 283
[5] Blasone M, Lambiase G and Luciano G G 2017 *Phys. Rev. D* **96** 025023
[6] Greiner W and Reinhardt J 1996 *Field quantization* Berlin Germany: Springer
[7] Oriti D 2000 ArXiv e-prints (Preprint gr-qc/0005010)
[8] Soffel M, Muller B and Greiner W 1980 *Phys. Rev. D* **22** 1935
[9] Unruh W G 1976 *Phys. Rev. D* **14** 870
[10] Blasone M, Lambiase G and Luciano G G In preparation

Mathematical aspects of an exactly solvable inflationary model

N Pintus[1] **and S Mignemi**[2]

[1] INAF-Osservatorio Astronomico di Cagliari, Via della Scienza 5, 09047 Selargius (CA), Italy
[2] Dipartimento di Matematica, Università di Cagliari, Viale Merello 92, 09123 Cagliari, Italy

E-mail: npintus@oa-cagliari.inaf.it

Abstract. The inflationary scenario is nowadays the most natural and simple way to solve the problems of the standard Friedmann-Lemaître-Robertson-Walker cosmology. Generally, inflation is driven by models of gravity minimally coupled to a scalar field ϕ rolling on a suitable potential $V(\phi)$. We discuss a particular model presenting two scenarios: the vacuum universe and the matter-filled universe, in order to find exact general isotropic and homogeneous cosmological solutions displaying an inflationary behavior at early times and a power-law expansion at late times.

1. Introduction

The inflationary scenario is based on the hypothesis that a period of accelerated expansion has taken place right after the Big Bang [1, 3]. It is today the widely adopted paradigm to interpret the high isotropy and homogeneity at large scales, the spatial flatness of our universe and the presence of about 24% of dark matter of unknown origin [4], and provides explanations of the formations of structures in the universe [5, 6].

The most common realization of inflation is based on models of gravity minimally coupled to a scalar field (the inflaton) with a self-interaction potential. These models can be classified according to the form of the potential [4]: in particular, among exponential potentials [7, 8] and even more general form [9, 10], double exponential potentials are the source of brane solutions of Einstein-scalar gravity, called domain walls, which can be analytically continued into Friedmann-Lemaître-Robertson-Walker cosmological solutions [11, 12] from the same model but with opposite scalar potential [13].

Inspired by this observation, we discuss the mathematical aspects of a model [12] of gravity minimally coupled to a scalar field ϕ defined by the action

$$I = \int \sqrt{-g} \left[\frac{R}{16\pi G} - 2 (\partial\phi)^2 - V(\phi) \right] d^4x \,, \tag{1}$$

where $R = g^{\rho\rho} R_{\rho\rho}$ is the trace of the Ricci tensor $R_{\mu\nu}$ and g is the determinant of the metric tensor $g_{\mu\nu}$. Indeed, in this model we use natural unities such that $c = \hbar = 1$. The scalar potential

$$V(\phi) = \frac{2\lambda^2}{3\gamma} \left(e^{2\sqrt{3}\beta\phi} - \beta^2 e^{2\sqrt{3}\phi/\beta} \right) \tag{2}$$

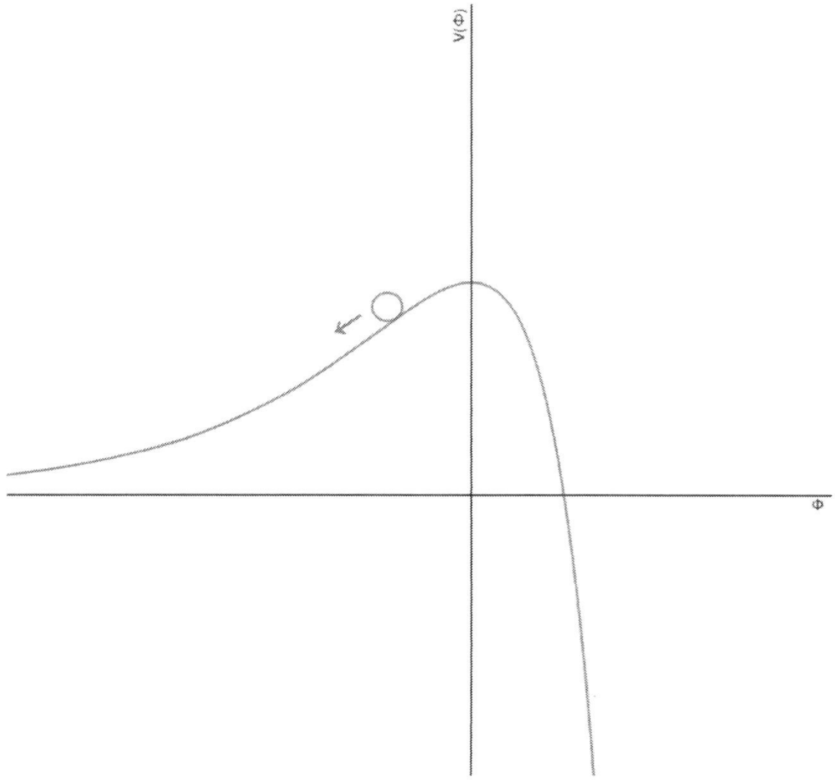

Figure 1. Plot of the potential (2) (in red) for $\lambda = 1$, $\beta = 1/\sqrt{3}$, with the inflaton (in blue) rolling on it.

depends on two parameters λ and β, with $\gamma = 1 - \beta^2$. The potential (2), for particular values of its parameters, is shown in Fig. 1. In general, the model is characterized by an exact solution in which the scale factor expands exponentially at early times and then behaves as a power law. Moreover, it admits a duality for $\beta \to \frac{1}{\beta}$ and for this reason we shall limit ourselves to the discussion of the case $0 < \beta^2 < 1$.

2. General solutions

We are interested in general isotropic and homogeneous cosmological solutions, with flat spatial sections, that we parametrize as

$$ds^2 = -e^{2a(\tau)}d\tau^2 + e^{2b(\tau)}d\Omega^2\,, \quad \phi = \phi(\tau)\,, \tag{3}$$

with τ a time variable.

With this parametrization, defining in the gauge $a = 3b$ two new variables $\Psi = a + \sqrt{3}\beta\phi$, $\chi = a + \frac{\sqrt{3}}{\beta}\phi$, we obtain the following exactly integrable vacuum equations

$$\ddot{\Psi} = \lambda^2 e^{2\Psi}\,, \qquad \ddot{\chi} = \lambda^2 e^{2\chi}\,, \qquad \dot{\Psi}^2 - \beta^2\dot{\chi}^2 = \lambda^2\left(e^{2\Psi} - \beta^2 e^{2\chi}\right)\,. \tag{4}$$

These equations are invariant under time reversal, $\tau \to -\tau$. The equations (4) admit first integrals

$$\dot{\Psi}^2 = \lambda^2 e^{2\Psi} + Q_1\,, \qquad \dot{\chi}^2 = \lambda^2 e^{2\chi} + Q_2 \tag{5}$$

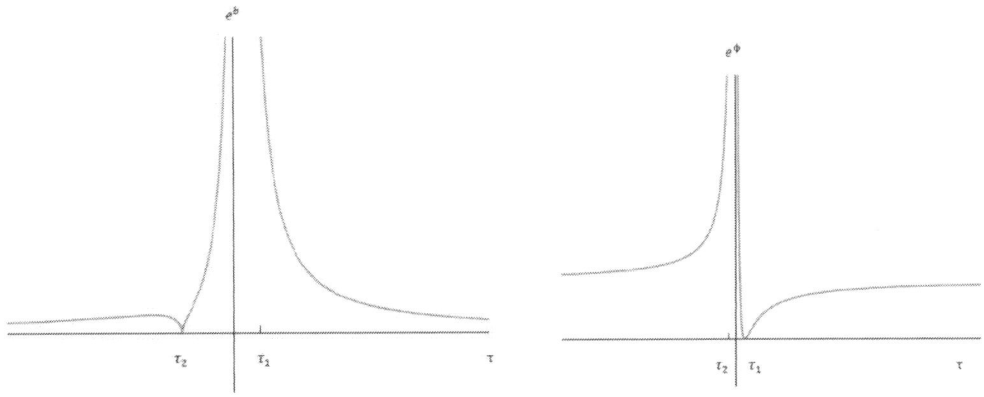

Figure 2. The solutions with $Q_i = 0$ for $e^{b(\tau)}$ and $e^{\phi(\tau)}$ when $\tau_1 \neq \tau_2$.

with Q_1 and Q_2 integration constants that satisfy the constraint $Q_1 = \beta^2 Q_2$. In this way, we can formally simplify the solutions of the system (4).

The solutions of the equations (4) are summarised in table 1, while the solution $Q_i = 0$ is depicted in Fig. 2.

Table 1. Solutions of the equations (4) depending on the sign of Q_i $(i = 1, 2)$, with $Q_1 = \beta^2 Q_2$ and $\tau_1 \neq \tau_2$ where τ_1, τ_2 are integration constants. We define $q^2 = Q_2 = Q_1/\beta$.

$Q_i > 0$	$Q_i =< 0$	$Q_i = 0$
$e^{2b} = \dfrac{q^2}{\lambda^2} \dfrac{\lvert \sinh\left[q\left(\tau - \tau_2\right)\right] \rvert^{2\beta^2/\gamma}}{\lvert \beta^{-1} \sinh\left[\beta q\left(\tau - \tau_1\right)\right] \rvert^{2/\gamma}}$	$e^{2b} = \dfrac{q^2}{\lambda^2} \dfrac{\lvert \sin\left[q\left(\tau - \tau_2\right)\right] \rvert^{2\beta^2/\gamma}}{\lvert \beta^{-1} \sin\left[\beta q\left(\tau - \tau_1\right)\right] \rvert^{2/\gamma}}$	$e^{2b} = \dfrac{1}{\lambda^2} \dfrac{\lvert \tau - \tau_2 \rvert^{2\beta^2/\gamma}}{\lvert \tau - \tau_1 \rvert^{2/\gamma}}$
$e^{2\sqrt{3}\phi/\beta} = \left\lvert \dfrac{\sinh\left[\beta q\left(\tau - \tau_1\right)\right]}{\beta \sinh\left[q\left(\tau - \tau_2\right)\right]} \right\rvert^{2/\gamma}$	$e^{2\sqrt{3}\phi/\beta} = \left\lvert \dfrac{\sin\left[\beta q\left(\tau - \tau_1\right)\right]}{\beta \sin\left[q\left(\tau - \tau_2\right)\right]} \right\rvert^{2/\gamma}$	$e^{2\sqrt{3}\phi/\beta} = \left\lvert \dfrac{\tau - \tau_1}{\tau - \tau_2} \right\rvert^{2/\gamma}$

We note that the parameter $q = \sqrt{Q_1}/\beta$ only sets the scale of the time parameter τ, while the physics interpretation is more transparent by changing the parametrization (3) defining the cosmic time t as $dt = \pm e^a d\tau$:

$$ds^2 = -dt^2 + e^{2b(t)} d\Omega^2 , \tag{6}$$

but the solutions in table 1 cannot be written in terms of elementary functions of t, except when $Q_i = 0$, $\tau_1 = \tau_2$. In this case, one obtains an expanding universe with $e^{2b} = e^{2\lambda(t-t_0)/3}$ and $\phi = 0$.

In the general case, to describe the smooth transition from inflationary expansion to a later power-law behavior, we consider those solutions that behave exponentially for $t \to -\infty$ and as a power law for $t \to \infty$. These are obtained for $Q_i = 0$, $\tau_1 \neq \tau_2$, as shown in Fig. 3, and correspond to an initial configuration where the scalar field is in the unstable equilibrium configuration at the top of the potential and then rolls down to $\phi = \infty$, that is $e^{2b} \sim e^{2\lambda t/3}$, $e^{2\sqrt{3}\phi/\beta} \sim$ const for $t \to -\infty$ and $e^{2b} \sim t^{2/3\beta^2}$, $e^{2\sqrt{3}\phi/\beta} \sim t^{-2/\beta^2}$ for $t \to \infty$.

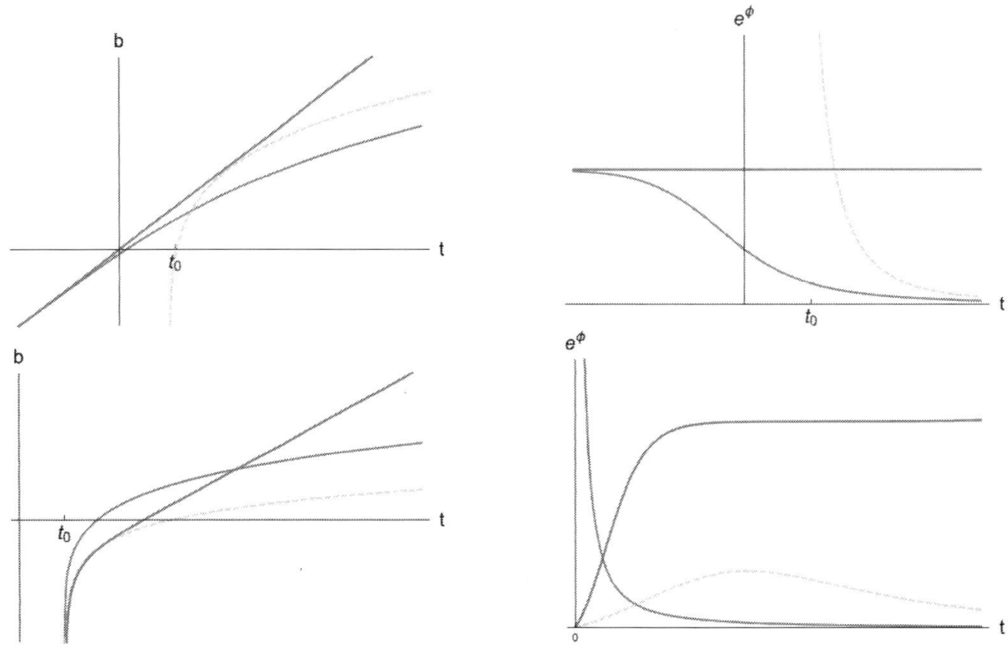

Figure 3. The solutions with $Q_i = 0$ (top panels) and with $Q_i \neq 0$ (bottom panels) in terms of the cosmic time t. The blue lines correspond to the solutions with $\tau_1 = \tau_2$ (top panels) and with $\tau_1 \neq \tau_2$ (bottom panels), while the continuous and the dashed curve (in red and green, respectively) are the expanding branches of the $\tau_1 = \tau_2$ (top panels) and of the $\tau_1 \neq \tau_2$ solutions (bottom panels).

3. Solutions with matter

Let us introduce matter in the form of a perfect fluid. The continuity equation $\dot{\rho} + (3p + \rho)\,\dot{a} = 0$, integrated for the equation of state $p = \omega\rho$ (with $\omega \geq -1$), gives $\rho = \rho_0\, e^{-(1+\omega)a}$, where ρ_0 is an integration constant. The Friedmann equations (4) become

$$\ddot{\Psi} = \lambda^2 e^{2\Psi} + \frac{3}{4}\rho_0\,(1-\omega)\,e^{(1-\omega)(\Psi-\beta^2\chi)/\gamma}\,, \quad \ddot{\chi} = \lambda^2 e^{2\chi} + \frac{3}{4}\rho_0\,(1-\omega)\,e^{(1-\omega)(\Psi-\beta^2\chi)/\gamma} \tag{7}$$

subject to the constraint

$$\dot{\Psi}^2 - \beta^2\dot{\chi}^2 = \lambda^2\left(e^{2\Psi} - \beta^2 e^{2\chi}\right) + \frac{3}{2}\rho_0\gamma e^{(1-\omega)(\Psi-\beta^2\chi)/\gamma}\,. \tag{8}$$

To investigate their properties, we put them in the form of a dynamical system, defining

$$X = \frac{1}{2\gamma\,(1-\omega)}\left(\dot{\psi} - \beta^2\dot{\chi}\right)\,, \quad Y = \dot{\chi}\,, \quad Z = \lambda e^{\chi}\,, \quad W = \sqrt{\lambda}e^{(\psi-\beta^2\chi)/2\gamma}\,. \tag{9}$$

With this choice of variables, we obtain a first-order differential system:

$$\dot{X} = \frac{\alpha}{2}W^2 - \frac{\beta^2}{2\gamma}Z^2 + \frac{1}{2\gamma}W^{4\gamma(1-\omega)}Z^{2\beta^2}\,, \qquad \dot{Y} = Z^2 + \alpha W^2\,, \qquad \dot{Z} = YZ\,, \tag{10}$$

Table 2. Classification of equilibrium points of dynamical system associated to the equations (7)

z^2	y	w^2	v^2	Eigenvalues	Kind of stability
0	$y_0 = 2\left(1 \pm \beta^{-1}\right)$	0	0	$(0, 0, y_0)$	unstable
0	2	$\frac{2(1-\omega)}{\alpha}$	0	$(\omega-1, \omega-1, 1-\omega)$	unstable
$\frac{4\gamma^2}{\beta^4}$	$-\frac{2\gamma}{\beta^2}$	0	0	$\left(\frac{2\gamma}{\beta^2}, \frac{2\gamma}{\beta^2}, \frac{4\gamma}{\beta^2}\right)$	stable
4	2	0	$\to 4$	$(-2, 2, -4)$	unstable
0	0	0	$\to 4\gamma^2$	$(-2\gamma, -2\gamma, -2\gamma)$	stable

with $\alpha = \frac{3\rho_0}{4\lambda}$ and W implicitly defined as

$$\frac{2\alpha}{1-\omega}W^2 + \frac{1}{\gamma}Z^{2\beta^2}W^{4\gamma(1-\omega)} = \frac{\beta^2}{\gamma}Z^2 + 4\gamma X^2 + 4\beta^2 XY - \beta^2 Y^2 . \tag{11}$$

The global properties of the solutions of the system (10) can be deduced from the study of its behavior near the critical points at finite distance or at infinity in the phase space.

In particular, the critical points at finite distance lie on two straight lines on the plane $Z = 0$, where $e^{2b} \sim t^{2/3}$, $e^{2\sqrt{3}\phi/\beta} \sim t^{2/\beta}$ as $t \to 0$.

The remaining critical points lie on the surface at infinity of the phase space. They can be studied by defining new variables $u = 1/X$, $y = Y/X$, $z = Z/X$, $w = W/X$ and considering the limit $u \to 0$. This is attained for $\tau \to \tau_0$, where τ_0 is a finite constant. For simplicity, we also define $v := z^{\beta^2}w^{2\gamma/(1-\omega)}u^{-\gamma(1+\omega)/(1-\omega)}$. One finds the critical points classified in table 2.

4. A more general form of the inflation potential

We discuss in this section a generalization of the potential (2) with a positive parameter $h \neq 1$, so that

$$V(\phi) = \frac{2\lambda^2}{3\gamma}\left(e^{2\sqrt{3}\beta h\phi} - \beta^2 e^{2\sqrt{3}h\phi/\beta}\right). \tag{12}$$

The action (1) and the parametrization (3) are unchanged. Defining two new variables $\tilde{\Psi} = \sqrt{3}\beta h\phi + 3b$, $\tilde{\chi} = \frac{\sqrt{3}h\phi}{\beta} + 3b$, the time-reversal invariant vacuum equations read

$$\ddot{\Psi} = \frac{\lambda^2}{\gamma}\left[\left(1 - h^2\beta^2\right)e^{2\Psi} - \beta^2\left(1 - h^2\right)e^{2\chi}\right], \quad \ddot{\chi} = \frac{\lambda^2}{\gamma}\left[\left(1 - h^2\right)e^{2\Psi} - \left(\beta^2 - h^2\right)e^{2\chi}\right] \tag{13}$$

with $\left(1 - h^2\beta^2\right)\beta^2\dot{\chi}^2 + \left(\beta^2 - h^2\right)\dot{\Psi}^2 - 2\beta^2\left(1 - h^2\right)\dot{\chi}\dot{\Psi} = \lambda^2 h^2\gamma\left(e^{2\Psi} - \beta^2 e^{2\chi}\right)$.

This system is no longer exactly integrable: as in the case of matter, we study the related dynamical system and we can find the critical points in table 3.

By comparing the $h \neq 1$ phase space with the one corresponding to $h = 1$, we see that the numerical behaviour of the solutions can be varied depending on the value of h. This allows the possibility of obtaining more realistic solutions for the expansion of the universe, as discussed in [14].

5. Conclusion

In order to describe explicitly the transition from the exponential inflation to the late time power-law expansion of the universe we have introduced an exactly solvable model of gravity

Table 3. Classification of equilibrium points of dynamical system associated to the equations (13)

z^2	y	w^2	Eigenvalues	Kind of stability
0	$y_0 = \frac{\beta^2-h^2}{\beta^2(1-h^2)}$	0	$(0, 0, y_0)$	unstable
0	y_0	$-\frac{\gamma}{\beta^2}y_0$	$\left(-y_0, -y_0, \frac{h^2\gamma}{\beta^2(1-h^2)}\right)$	unstable
$\frac{\gamma}{1-h^2\beta^2}$	$\frac{1-h^2}{1-\beta^2h^2}$	0	$\left(-1, -1, -\frac{2}{\gamma}\left(1-h^2\beta^2\right)\right)$	stable
1	1	1	$\left(-1, -1, -\frac{2}{\gamma}\left(1-h^2\beta^2\right)\right)$	unstable

minimally coupled to a scalar field subject to a doubly-exponential potential, whose solution corresponds to the decay from an initial configuration where the scalar field is in an unstable vacuum state. The aim of our mathematical investigation has been to study the general solutions of such model in vacuum and in the presence of matter. It turns out that the exponential solution corresponds to very specific initial conditions, while generic solutions present instead power-law inflation. The physical relevance of the solutions depends on the possibility that the inflationary mechanism provided by our model gives correct predictions on the evolution of perturbations and on observable cosmological data. This is possible if one generalizes the potential as in (12). Although in the general case it is not possible to find exact solutions, the behavior of the solutions is qualitatively similar to the previous case, except that the new parameter allows for the possibility of obtaining more realistic predictions. Some exact solutions however are still available for specific initial conditions, which essentially coincide with those found in the previous case.

References

[1] Starobinski A 1980 *Phys. Lett.* B **91** 99
[2] Guth A 1981 *Phys. Rev.* D **23** 347
[3] Linde A 1982 *Phys. Lett.* B **108** 389
[4] Fré P, Sagnotti A and Sorin A 2013 *Nucl. Phys.* B **877** 1028
[5] Lyth D and Liddle A 2009 *The Primordial Density Perturbation* (Cambridge: Cambridge University Press)
[6] Bassett B, Tsujikawa S and Wands D 2006 *Rev. Mod. Phys.* **78** 537
[7] Lucchin F and Matarrese S 1985 *Phys. Rev.* D **32** 1316
[8] Salopek D and Bond J 1990 *Phys. Rev.* D **32** 3936
[9] Ellis G and Madsen M 1991 *Class. Quantum Grav.* **8** 667
[10] Chervon S, Zhuravlev V and Shchigolev V 1997 *Phys. Lett.* B **398** 269
[11] Cadoni M and Ciulu M 2014 *J. High Energy Phys.* JHEP05(2014)089
[12] Mignemi S and Pintus N 2015 *Gen. Relativ. Grav.* **47** 51
[13] Skenderis K and Townsend P 2006 *Phys. Rev. Lett.* **96** 191301
[14] Cadoni M, Franzin E and Mignemi S 2016 *Eur. Phys. J.* C **76** 483